BASICS OF PLANT PHYSIOLOGY

BASICS OF
PLANT PHYSIOLOGY

Nirmal Chandra Pradhan

ANMOL PUBLICATIONS PVT. LTD.
NEW DELHI - 110 002 (INDIA)

ANMOL PUBLICATIONS PVT. LTD.
H.O.: 4374/4B, Ansari Road, Darya Ganj,
New Delhi-110 002 (India)
Ph.: 23278000, 23261597
B.O.: No. 1015, Ist Main Road, BSK IIIrd Stage
IIIrd Phase, IIIrd Block,
Bangalore - 560 085 (India)
Visit us at: www.anmolpublications.com

Basics of Plant Physiology

First Published, 2008
ISBN 978-81-261-3529-5

PRINTED IN INDIA

Printed at Mehra Offset Press, Delhi.

Contents

Preface

This book has developed largely from the courses in plant physiology. The authors have attempted to organize a discussion of the fundamental facts and principles of this subject which can be included between the covers of a volume of moderate dimensions. It is believed that this book can be readily adapted to use with any introductory course in plant physiology based upon prerequisites of general botany and general chemistry.

The authors believe that effective teaching in any science not only should lead to the acquisition by the student of the basic facts, principles, and view-points of that science, but also should train the student in the use of that science in the interpretation of natural phenomena. Students should come to think of plant physiology not as a subject but as a useful tool which canbe used in the explanation of plant behavior under natural or cultural conditions. Frequent exposure to properly selected problems and discussion ques-tions will usually aid the student to use the facts and principles of a science as well as to learn them.

Author,

Chapter 1

The Field of Plant Physiology

In order to understand the mechanism of a plant in its entirety many separate phases of the dynamic activity of plants have been recognized and studied as individual processes. In his efforts to interpret these individual processes and their interrelationships the plant physiologist is confronted with many problems: What is the mechanism by which water, gases, and solutes enter a plant from its environment? How do such substances pass out of a plant into its surroundings? How are foods and other complex organic compounds synthesized in the plant? How are they utilized in the development and maintenance of plants as living systems? What transformations of energy occur within a plant, and what exchanges of energy take place between a plant and its environment? How are water and solutes transported from one part of a plant to another? How are new tissues constructed? How is the development of one organ or tissue coordinated with the development of other organs or tissues? Why does a plant produce only vegetative organs at certain stages in its life cycle and reproductive organs only at other stages? How are individual plant processes and the development of a plant as a whole influenced by environmental conditions? All of these problems and many other related or subsidiary ones lie within the province of the branch of science which is known as plant physiology.

For convenience, the study of plant life is subdivided into various branches such as physiology, morphology, anatomy, ecology, pathology, genetics, etc. Such a classification is necessarily more or less arbitrary. Neither physiology nor any other phase of plant life can be singled out for study without some consideration of plants from other viewpoints. A particularly intimate interrelationship exists between the structures and processes of plants. Every physiological process is conditioned by the anatomical arrangement of the tissues, and by the size, configuration, and other structural features of the cells in which it occurs. Furthermore, the coordinated development of cells and tissues, i.e., of the plant itself, is a complex of physiological processes. Thus the sciences of plant physiology and plant anatomy merge in the study of plant growth.

Just as different species of plants differ in outward configuration and internal anatomy, so do they also differ in physiology. The world of plants includes a great number of very diverse kinds of organisms that range in size from simple bacterial cells a few microns long to the enormous redwoods of our Pacific coast and mountain forests. Their difference in size, although striking enough, is not as fundamental as another distinction which exists between redwoods and bacteria. The redwoods and all other green plants are able to manufacture their own food, while the bacteria (with a very few exceptions) and all other non-green plants must obtain their nourishment from some outside source. The physiology of the green and the non-green plants is therefore basically unlike. This book is essentially a discussion of the physiology of the chlorophyllous (green) plants with the emphasis on the vascular green plants. The physiological processes of the bacteria and fungi have been considered only when they have a direct bearing on the physiology of the green plants.

With few exceptions all vascular green plants carry on the same fundamental physiological processes, but there are many differences in subsidiary processes. Most green plants, for example, synthesize starch, but many don't. Numerous similar variations occur in the kinds of metabolic products built up by various species of plants. Most of the physiological differences among the species of green plants, however, are quantitative rather than qualitative. All chlorophyllous plants carry on photosynthesis, but when different species are exposed to the same environmental conditions the rate at which the process takes place may differ greatly from species to species. A similar situation holds with respect to practically all of the other processes occurring in green plants. Even varieties of the same species often exhibit marked differences in physiological behaviour when exposed to a given environmental complex. Some varieties of wheat, for example, are markedly cold resistant while others are not.

THE RELATION OF PHYSIOLOGY TO THE PHYSICAL SCIENCES

Formerly the opinion was almost universal that living organisms owe their distinctive properties to the possession of subtle and unknown forces which are peculiar to living matter. At the present time such vitalistic theories find very few advocates. The contrary and now widely held assumption is that living organisms operate in accordance with the same physico-chemical principles that hold in the inanimate world. The complexity and elusiveness of living processes are not assumed to be due to intangible unknown varieties of energy but to the interplay of recognizable physico-chemical forces in the complex organized system of the protoplasm. Adoption of this latter point of view has led to a widespread use of the tools of physics and chemistry in experimental work on plants, and to the interpretation of plant processes in terms of these two sciences. This has led to notable progress in our understanding of the physiology of plants and has permitted the analysis and expression of many physiological relations in quantitative terms.

A knowledge of certain fundamental principles of physics and chemistry is therefore essential to the understanding of physiological processes. For this reason several of the earlier and parts of some of the later chapters in this book are devoted to a brief exposition of those underlying principles of the physical sciences with which a student of plant physiology should be familiar.

THE RELATION OF PLANT PHYSIOLOGY TO THE AGRICULTURAL SCIENCES

Green plants are not only the ultimate source of all food but supply the raw materials for many of our basic industries. With the rise of modern industrial civilization both the quantities and kinds of plant products which we utilize have increased rather than decreased. In addition to foods some of the more important raw products obtained from plants are wood, textile fibers, pulp, rubber, vegetable oils, gums and drugs. Even most of the so-called synthetic products of the chemist are not synthetic in the sense that they have been built up from simple inorganic compounds, but only insofar as they represent modifications of naturally occurring plant products.

An industrial civilization not only requires a wide variety of plant products but insists that these products meet certain standards of quality. The successful cultivation of plants has, therefore, become a highly skilled occupation, and the agricultural sciences are rapidly becoming a domain of the specialist. Success in controlling the activities of living plants can not be achieved without some understanding of the processes which occur within them, and of the effects of environmental conditions upon these processes. The problems of the forester, the fruit grower, the cotton planter, the floriculturist, the grain farmer, and of all others who cultivate plants differ in detail, yet they all have a fundamental similarity in requiring an application of the principles of plant physiology for their solution.

Fundamental investigations in plant physiology have contributed in many ways to improved methods of propagating, cultivating and harvesting economically important plants, and to methods of handling and storing many plant products. Furthermore, control of the fungous diseases and insect predators of plants often requires application of the principles of plant physiology. Much of the investigational work carried on by scientific agronomists, horticulturists, floriculturists, and foresters actually lies in the field of pure or applied plant physiology, although often it is not formally classed as such.

PLANT PHYSIOLOGY AS A SCIENCE

The origin of mans first conscious interest in plants long antedates recorded history. Agriculture had already become a highly developed art thousands of years before any experimental study of plant processes began. Consequently, there had grown up a vast body of traditional plant lore which passed orally from father to son, generation after generation. This practical knowledge of plants developed over many centuries as a result of countless, mostly involuntary trial and error experiences and innumerable observations of plant behaviour under all kinds of circumstances. Which of this mass of mingled facts and beliefs regarding plants is alive in the consciousness of the common man today. The closer he lives to the soil and the more familiar he is with traditional plant lore, the more it influences his thoughts and actions. Many of these customarily accepted beliefs are essentially sound and most of them contain elements of truth. Others, however, are entirely erroneous, and not a few are tempered with superstitions, so me of which have an unbroken lineage back to the days of witch-doctors and savagery. No reputable botanist has held for generations, for example, that plants obtain their food from the soil, yet this and many other fallacious beliefs are still widely entertained among the general population. The value of practical information about plants should not be underrated, since its perpetuation in the mind of man permitted the development of agriculture to a high plane as a practical art before any widespread investigation of plants from a scientific point of view was undertaken. Nevertheless, traditional plant lore is not only often inadequate, but is riddled with misconceptions, and often suffers from points of view which are inherently stultifying to the acquisition of further knowledge.

The layman, for example, often personifies plants in an attempt to explain their behaviour. Man has desires and foresight and it is often assumed, either consciously or tacitly, that plants are similarly endowed. To many, for example, the statement that roots grow downwards in search of water, or that stems grow upwards in order to reach the light are accepted as adequate explanations of plant behaviour. Mans knowledge that water and light are essential to plants is not evidence that plants are similarly aware of these facts. To assume that plants realise their needs and are able to act in conformity with their requirements is equivalent to crediting them with a high order of intelligence. Explanations of plant behaviour are commonly encountered in which purposeful action on the part of plants is tacitly or deliberately implied, although there is no justification for the adoption of such a point of view. Furthermore, the layman seldom pursues his quest for information about plants beyond the stage of observation, while the scientist frequently does.

Observation has suggested, for example, that light is necessary for the continued existence of plants. To one who is scientifically minded, either by instinct or training, the obvious next step is to test this postulate experimentally. If the suggested hypothesis is substantiated by experiment it is tentatively accepted as a theory. Theories such as those proposed in explanation of the processes or reactions of plants, together with the experimental results which are considered to support them, are usually published in a scientific journal or monograph and thus exposed to evaluation and further experimental testing by other scientists.

Continued experimentation may lead to substantiation, rejection, or modification of the theory as originally proposed. Modification would undoubtedly be the fate of the hypothesis used as an example, since sooner or later some investigator would find that non-green plants can thrive in the absence of light. Experiments often raise more questions than they answer. New approaches to the problem under consideration as well as desirable new lines of inquiry are constantly opening up to the alert investigator. In this way experimentation leads to more experimentation, more facts accumulate, and more theories are proposed. Some of the suggested hypotheses are confirmed, others are rejected, and still others are modified. Most of them, sound or fallacious, in turn suggest further observation and experimentation. As a result of such endless and painstaking labours there is slowly built up that vast, complex, and ever changing body of knowledge which we refer to as a science. The system of subjecting all hypothetical explanations of natural phenomena to experimentation is the essence of the scientific method. It is the characteristic methodology of all sciences. Progressive modification of accepted concepts in the light of new experimental findings continually increases the soundness of scientific generalizations. Thus there are incorporated into any science theories and generalizations in various stages of acceptance. Some stand upon such a firm substructure of facts that they are accepted by all authorities in the field. Others, less securely supported by experimental results, are subscribed to by some but rejected by other workers. Finally, in any science there are always some theories which are so dubious that they find only a few advocates. Furthermore, some of the theories now widely held sooner or later will be discarded as a result of new findings or as a consequence of different interpretations of facts already known.

However, not all scientists are in agreement regarding the interpretation of the same sets of facts. Although this state of affairs is entirely consistent with the spirit of scientific research, it is frequently puzzling to students and laymen. Differences of opinion regarding the hypotheses which suitably explain scientific phenomena are most likely to arise when experimental data are inadequate. Disagreements regarding the interpretation of experiments and observations are often inevitable steps in the clarification of scientific generalizations. Controversies usually focus attention upon gaps in our factual information. Frequently, therefore, they are stimulating to research, and often lead to a further enrichment of human knowledge. Without the knowledge of plant processes which has been slowly accumulated through more than two centuries of observation, experimentation, and critical evaluation by numerous workers in all parts of the world, this book could never have been written. In spite of the patient labours of these many workers vast gaps still exist in our understanding of the physiology of plants gaps which are reflected in the necessarily inadequate treatment of many topics in this book. The future of this science and all others lies in the hands of the front-line investigators who wage a continual struggle for the extension of human knowledge.

Chapter 2

Properties of Solutions

Water is the most abundant compound present in all physiologically active plant cells. The water which occurs in a liquid state in plant cells invariably contains other substances dissolved in it and usually also contains dispersed particles which are not in true solution. When the particles dispersed throughout the water are within a certain range of sizes the system of water plus particles falls into the category of colloidal systems. The complicated dynamics of living systems can be largely interpreted in terms of the physicochemical properties of solutions and colloidal systems, one component of which is water, although it should not be inferred that non-aqueous solutions and colloidal systems are entirely absent in living organisms. Similarly liquid water never occurs in the pure state in the natural environment of living organisms. The water of streams, lakes, and oceans invariably contains various substances in solution and usually in the form of dispersed particles as well. This is likewise true of the soil water. Even raindrops, the products of natural distillation, contain gases which have dissolved in them from the atmosphere.

GENERAL NATURE OF SOLUTIONS

Simple solutions are systems in which one component (the solute) is dispersed throughout the other (the solvent) in the form of molecules or molecules and ions. Theoretically the solvent may be a gas, a solid, or a liquid, but solutions in which the solvent is a liquid are by far the most important in living organisms. Except in extremely concentrated solutions the average distance between the solute particles is usually very great relative to their size. Naturally occurring solutions, whether in living organisms or their environment, usually contain a number of different solutes and are often exceedingly complex. Water is the commonest and most important of all solvents both in the inorganic world and in the realm of living organisms. The further discussion will be principally in terms of aqueous solutions.

SOLUTIONS OF A GAS IN A LIQUID

The water present in living organisms usually contains dissolved gases. Those most commonly present are carbon dioxide, oxygen, and nitrogen. Practically all the water in the environment of organisms in oceans, lakes, streams, soils, and raindrops also contains these gases in solution, and sometimes others as well. A given volume of water or any other liquid will hold only a limited quantity of gas in solution at a given temperature. When no more of a certain gas can be dissolved in a liquid it is said to be saturated with respect to that gas. Gases vary widely in their solubility in water, but in general fall in to two groups: those which are sparingly soluble and those which are highly soluble. When only a small fraction of a unit volume of a gas will dissolve in a unit volume of water the gas is classified in the sparingly soluble group. When from one to many unit volumes of a gas will dissolve in one unit volume of water it is classified in the highly soluble group. Oxygen, hydrogen, and nitrogen are familiar examples of gases belonging to the first group, while carbon dioxide, ammonia, and hydrogen chloride are examples of the second. When gases are highly soluble in water it is usually evidence that a chemical reaction takes place between the gas and water. There actions between water and carbon dioxide and water and ammonia are indicated in the following equations:

$$CO_2 + H_20 \rightarrow H_2C0_3NH_3 + H_20+1NH_4OH$$

In a solution of such gases not only are molecules of the gas present, but also molecules of a compound formed by the reaction of the gas with water. The apparently great solubility of gases such as these is due to the formation of relatively soluble compounds by the reaction of the gas with the water.

Increase in pressure of a gas above a liquid increases the solubility, i.e. the concentration of that gas in the liquid. For sparingly soluble gases the increase in solubility is directly proportional to the increase in pressure (Law of Henry). This principle also holds qualitatively, but not quantitatively, for the very soluble gases.

When a mixture of several gases is maintained over water, each dissolves independently and in accordance with the gaseous pressure (partial pressure) which it exerts against the surface of the water (Law of Dalton). This principle holds strictly only for those gases which are slightly soluble in water. For example about one-fifth of the atmospheric pressure is due to the oxygen present. Assuming an atmospheric pressure of 760 mm. Hg (the value at standard conditions), the pressure due to the oxygen is equivalent to about 152 mm. Hg. The quantity of oxygen which dissolves in water exposed to the air is the same as that which would dissolve if oxygen only at a pressure of 152 mm. Hg occupied the space over the water.

SOLUTIONS OF A LIQUID IN A LIQUID

In general solutions of a liquid in a liquid fall into two classes: those in which the liquids are freely miscible with each other in all proportions and those in which each liquid reaches a definite point of saturation in the other. Alcohol and water, for example, mix with each other in all proportions; such a solution is an example of the first mentioned type. Many oily liquids also are miscible with each other in all proportions. A familiar example is the solution of lubricating oil in gasoline. In solutions of this kind the liquid present in excess is usually considered to be the solvent. In a 50 per cent solution of alcohol and water, either liquid could be considered the solvent, or either could be considered the solute. Ether, chloroform, and many other liquids are sparingly soluble in water. After water and ether, for example, are shaken together in a flask, two distinct layers of liquid separate upon standing. The upper layer consists of the lighter ether, saturated with water, while the lower consists of water saturated with ether. In both layers, however, the concentration of solute present at saturation is small. In the upper layer ether is the solvent, water the solute; the converse is true of the lower layer.

SOLUTIONS OF A SOLID IN A LIQUID

This is by far the most familiar type of solution and in many respects the most important. Substances in the solid state vary greatly in their solubility in water, ranging all the way from those which are virtually insoluble to those which are extremely soluble. There is usually a limit to the amount of any solute which can be dissolved in a given volume of water at a given temperature. When this limiting concentration is reached the solution is said to be saturated, following the same terminology used with solutions of gases and liquids in liquids. It is almost impossible to prepare a true saturated solution of any solute unless some of it is also present in the solid state. Under certain conditions, and only when none of the undissolved solid is present, a super-saturated solution can be prepared; that is, a solution in which the concentration of the solute is greater than that in the saturated solution. If a fragment of the solid solute be added to such a solution the excess solute usually crystallizes out immediately, and the concentration of the solution decreases to that usually present at saturation. Usually increase in temperature increases the solubility of a solid in a liquid but there are some exceptions to this principle. The solubility of common salt (NaCl) in water, for example, is only slightly influenced by temperature, while the solubility of many calcium compounds is decreased by an increase in the temperature of water.

METHODS OF EXPRESSING THE COMPOSITION OF SOLUTIONS

If a mol of any soluble compound be dissolved in just enough water to make exactly one litre of solution the result is a volume molar solution. Since the volume of water changes with temperature it is usual to specify that the solution is to be made up to a litre volume at 200°C. Whenever the word molar is used without qualification this type of solution is meant. Similarly, if half the molar weight, or one-tenth the molar weight of a substance be dissolved in enough water to make exactly one liter of solution the result is a half molar (0.5 IVI) or tenth molar (o.i IVI) solution, respectively, etc. Gram molar weights of all substances contain the same number of molecules. Estimates made by various methods show this number to be about 6.o6 × 1023 molecules. Hence one liter of any volume molar solution will contain this number of solute molecules, one millilitre (0.001 liter) will contain one thousandth of this number, and so on. Equal volumes of all solutions of the same volume molarity contain the same number of solute molecules but different numbers of solvent molecules. If a given volume of a volume molar solution be diluted with an equal volume of water, the result is a 0.5M solution; if a given volume be diluted with nine volumes of water the result is a 0.1M solution etc. Therefore a volume molar solution of any strength may be diluted with water and the resulting more dilute solution will have a volume molarity in proportion as it has been diluted.

If a mol of any soluble substance be completely dissolved in 1000g. ofwater the result is a weight molar solution. Such solutions are often called molal solutions. They are used principally in experimental work upon various osmotic phenonema.The addition of a mol of most solids to a liter of water will increase the volume of the resulting solution to more than one liter. This increase in volume is called the solution volume of the solute. The solution A mol is the molecular weight of a compound in grams.

There is a shrinkage in volume when the solute is added to the solvent. On the other hand the solution volume of some compounds, especially the sugars, is considerable. When a mol of sucrose is added to 1000g. of water the resulting solution will have a volume of 1207cc. at 0^0C. Hence the solution volume of sucrose is 207cc. The solution volume of a mol of sodium chloride, on the other hand, is only about 18cc. Since every solute has a different solution volume, it follows that equal volumes of weight molar solutions do not contain the same number of either solvent or solute molecules.Neither will the dilution of a given volume of a weight molar solution with an equal volume of water result in a 0.5 molal solution. In other words the concentration of a weight molar solution does not change in proportion to the amount by which it has been diluted. This is a fact which is sometimes overlooked in experimental work.

In physiological work it is often convenient to make up solutions on a percentage basis. Such solutions are made up either on the basis of percentage by weight, or percentage by volume. Solutions of solids in water or other solvents are made up on a weight percentage basis. A 10 per cent sodium chloride solution, for example, is made by dissolving 10 g. of sodium chloride in 90 g. of water. Solutions of liquids in water or other solvents can also be prepared on a weight percentage basis. It is simpler, however, to make up such solutions on a volume percentage basis. On this basis a 10 per cent solution of alcohol is prepared by adding 10cc. of alcohol to 90cc. of water. The percentage system of solutions has no direct relation to either the volume molar or weight molar systems.

ELECTROLYTES AND NON-ELECTROLYTES

Some aqueous solutions readily conduct an electric current; others do not. The former are called electrolytes; the latter non-electrolytes.The solutions of all acids, bases, and salts are electrolytes. Solutions of most organic compounds such as the sugars, alcohols, ketones, and ethers are non-electrolytes.

This process is called electrolysis. If hydrochloric acid is the electrolyte, for example, hydrogen gas will be liberated at the negative pole (cathode) and chlorine gas will be liberated at the positive pole (anode).If electrolysis is continued long enough eventually all of the hydrochloric acid present in the system will be decomposed into hydrogen gas and chlorine gas. Strictly speaking the term electrolyte refers only to a solution of an ionized substance, but it is also often applied to any compound which, when dissolved in water, produces ions.A similar dual usage of the term non-electrolyte also prevails.

The occurrence of electrolysis, as well as the unusually large effects of electrolytes on the osmotic pressures of solutions, have led to an explanation of the behaviour of electrolytes in terms of the theory of electrolytic dissociation. This theory was first advanced by the Swedish chemist Arrhenius who introduced it in essentially its present form in 1887. According to the Arrhenius theory when an electrolyte is dissolved in water some of the molecules dissociate into two kinds of particles, one positively charged, the other negatively charged. Each of these particles is called an ion. Ions are supposed to be present in a solution whether any electrolysis occurs or not.Two, three, four or even more ions may be formed from a single molecule.The conduction of an electrical current by an electrolyte is due to the presence of these ions.

The positively charged ions, which in electrolysis migrate towards the cathode, are called cations; the negatively charged ions which migrate toward the anode are called anions. A cation may carry one, two, three, or even four positive charges; an anion may carry from one to several negative charges.When an electrolyte dissociates the number of positive charges carried on thecations is always equal to the number of negative charges carried by the anions.

The Arrhenius theory does not assume complete ionization of all of the solute molecules in an electrolyte but that the molecules are continuously dissociating into ions, while free ions in the solution are continuously reuniting and forming molecules, both processes proceeding at an equal rate whenever an equilibrium condition prevails.An equilibrium of this sort, which is maintained by two opposing processes proceeding at equal rates is termed adynamic equilibrium. Electrolytes vary greatly in degree of dissociation. Those in which a large proportion of the molecules are maintained in a dissociated condition are termed strong electrolytes; those of which the contrary is true, weak electrolytes.Strong electrolyte solutions conduct electric currents better than weak electrolyte solutions of equal concentration. In general the more dilute a given electrolyte solution the larger the proportion of dissociated molecules present. In extremely dilute solutions ionization is practically complete. Increase in temperature reduces the dissociation of an electrolyte.

It is probable that many of the so-called non-electrolytes also dissociate very slightly in solution, but the degree of dissociation of such compounds is so small that it can be detected only by very refined methods, if at all. Even water, as the subsequent discussion will show, dissociates slightly, producing hydrogen and hydroxyl ions.

ACIDS, BASES, AND SALTS

An acid may be defined as a substance which produces hydrogen ions (H^+) when dissolved in water. The characteristic properties of acids are due to the hydrogen ions produced. The most common in Organic acids are hydrochloric (HCl), nitric(HNO_3), and sulfuric($H_{25}0_4$). In living organisms a large group of more complex, but much weaker acids, known as organic acids, play important roles. The strength of an acid depends upon its degree of ionization; the greater the proportion of hydrogen ions an acid produces in solution at a given concentration, the stronger it is.

A base may be defined as a substance which produces hydroxyl ions (OH^-) when dissolved in water. The characteristic properties of bases are due to the hydroxyl ions produced. Some of the commonest bases are sodium hydroxide (NaOH), calcium hydroxide ($Ca(OH)_2$), and potassium hydroxide (KOH). The strength of a base, like that of an acid, depends upon its degree of ionization. The greater the proportion of hydroxyl ions a base produces in solution at a given concentration, the stronger it is.

The theory of electrolytic dissociation accounts much more satisfactorily for the behaviour of weak electrolytes than of strong electrolytes. For this and other reasons Debye and Hiickel in 1923, and other modern investigators, have postulatedthat strong electrolytes, at least, are completely dissociated. The fact that they behave as if only partly dissociated is accounted for in terms of inter ionic attractions. These attractions are supposed to exert what may be roughly described as a braking effect upon the movement of the individual ions.This prevents them from operating at their maximum effectiveness in conducting an electric current or in other phenomena depending upon ionic mobilities. The greater the concentration of an electrolyte, the closer together are the ions, and, according to this view, the less their apparent dissociation.

A salt may be defined as a compound which has been produced by the union of an anion or anions of an acid with the cation or cations of a base. Salts are produced when an acid and a base are brought together in a solution as a result of a chemical union between the hydrogen ion (s) of the acid and the hydroxyl ion (s) of the base, forming water. This reaction is called neutralization.

Since water is practically undissociated, neutralization reactions go rapidly to completion. The reverse reaction, indicated in the above equations by the arrows pointing to the left, is called hydrolysis. Under certain conditions hydrolytic reactions may proceed at a rapid rate, although in the examples presented above the speed cf the reaction towards the left (hydrolysis) is negligible compared with the speed of the reaction towards the right (neutralization).

NORMAL SOLUTIONS

The concentrations of acids and bases are most commonly expressed in terms of normal solutions. A normal solution of an electrolyte contains in a dissolved state per liter of solution at 200 C. a weight of the compound in grams equal to its molar weight divided by its hydrogen equivalent. The hydrogen equivalent of a compound is defined as the number of replaceable hydrogen atoms in one molecule, or the number of atoms of hydrogen with which one molecule could react.

Thus a normal solution of an acid contains 1.008 g. of replaceable hydrogen per liter; a normal solution of a base 17.008 g. of replaceable hydroxyl per liter. By this system the concentration of any acid, base, or salt can be designated as 0.1 N, 0.5 N, 2 N etc., as the case may be.

The normality of an acid solution is a measure of its total acidity, i. e.of its concentration in terms of replaceable hydrogen ions. Similarly the normality of a base solution is a measure of its total basicity. Since 1.008 g.of replaceable hydrogen represents the same number of ions as 17.008 g. of replaceable hydroxyl (why?) it is evident that equal volumes of acid and base solutions of equal normality will exactly neutralize each other. Hydrogen Ion Concentration. The effects of acids upon chemical reactions and upon physico-chemical conditions generally in both inorganic systems and in living organisms are due principally to the hydrogen ions which they produce when in solution.

Some of the most fundamental of physiological phenomena are markedly influenced by the concentration of hydrogen ions in the medium in which they occur. For many purposes therefore it is more important to have some sort of a measuring stick of the concentration of hydrogen ions present in a solution than of the total acidity of the solution.Total acidity, as already noted, is customarily expressed in terms of normality. It is also entirely possible to use the normal system for expressing the concentration of hydrogen ions. In a normal solution of hydrochloric acid, for example, about 78 per cent of the molecules are dissociated. The term normal as used in the preceding sentence refers to total acidity, that is, to all the ionizable hydrogen whether actually present as ions or combined with anions in the form of molecules. In terms of the hydrogen ions present, however, such a solution is only 0.78 N. The term normal as used in this latter sense refers only to the ionized hydrogen. We may speak therefore of a normal solution of hydrogen ions as well as of a normal solution of an acid.Since no acid is ever completely dissociated, a normal solution of any acid will always be less than normal when its concentration is expressed in terms of the hydrogen ions present.

In order to prepare a normal solution of hydrogen ions it is necessary to make up a solution which is more than normal in terms of total acidity. Such a solution must be of the precise strength and degree of dissociation that exactly i.008 g. of the ionizable hydrogen present are actually in the dissociated form as ions per liter of the solution. Although hydrogen ion concentration can be readily expressed in terms of normalities, in actual practice this system is not generally used because it is apt to prove cumbersome, especially when it is necessary to refer to the very small concentrations of hydrogen ions usually dealt with in biological problems.

The hydrogen ion concentration of a solution is now quite generally defined in terms of its pH value. The pH value bears a simple mathematical relation to the hydrogen ion concentration of a solution in terms of its normality. Because of the practically universal acceptance of this system it is necessary to understand the significance of the term pH and its relation to hydrogen ion concentration expressed in terms of normality. The relation between pH and hydrogen ion concentration in terms of normality is a logarithmic one. The pH of a normal solution of hydrogen ions is 0, of a 0.1 N solution i, of a 0.01 N solution 2, etc. Zero is the logarithm of i, i is the logarithm of 10, 2 is the logarithm of 100, etc. The pH value for any solution is the negative of the logarithm of the hydrogen ion concentration in terms of normality. It may also be defined as the logarithm of the number of liters of solution that contains one gram atomic weight of hydrogen ions.

All aqueous solutions as well as pure water contain hydrogen ions in some concentration. Corresponding to the concentration of hydrogen ions is a certain definite concentration of hydroxyl ions. It can be shown by the principle of mass action that the mathematical product of the concentration of hydroxyl ions and the concentration of hydrogen ions in a solution is a constant. This may be expressed as follows:

(H+) × (0H)K.Kio14 at 2O C.

Hence as the pH of a solution is increased the pOH decreases and vice versa. For example, if the pH increases from 5 to 6 the poll decreases from 9 to 8. Furthermore, only at pH 7 can the concentration of hydrogen ions equal the concentration of hydroxyl ions. This is therefore the neutral point on the pH scale and corresponds to the dissociation of pure water. This pH value represents a dissociation of only one water molecule in approximately every 555,000,000.

Values below 7 on the pH scale represent the acid range, those above 7 the alkaline range of the scale. An acid solution is one with a larger concentration of hydrogen ions than hydroxyl ions, while in an alkaline solution the reverse is true. The lower the pH value the greater the hydrogen ion concentration of a solution. A pH value of 5 represents ten times the hydrogen ion concentration of a solution with a p11 of 6 and one hundred times the hydrogen ion concentration of one with a ph value of 7, etc. This is due to the fact, previously emphasized, that the numbers on the pH scale are related to each other as logarithms and not as ordinary arithmetic numbers.

The hydroxyl ion concentration of a solution could be expressed in p011 units instead of in pH units. For alkaline solutions especially this would seemto be a logical practice. But because of the definite mathematical relationship between the pH and pOH the p11 value alone also defines the p011 value. Hence both the acidity of a solution in terms of H + ions, and its alkalinity in terms of OH ions may be, and customarily are, expressed in terms of pH units. It is possible for aqueous solutions to exist which have a pH value of less than 0 (i.e. a minus pH value) or more than 14. A N solution of hydrochloric acid, for example, would have a pH value of a little less than o; correspondingly a N solution of sodium hydroxide would have a pH value a little greater than 14. In actual experience, however, solutions of such extreme concentrations are seldom encountered that the sodium acetate solution retards in some way changes in pH value upon the addition of an acid while a solution of sodium chloride does not.

Solutions of such compounds as sodium acetate which are relatively resistant to changes in p11 due to the addition or loss of hydrogen or hydroxyl ions are known as buffer solutions, and this property of solutions is called buffer action. Solutions such as sodium chloride which show no buffer effect are called unbuffered solutions.If 1 cc. of a normal sodium hydroxide solution be added to 10 cc. of a normal solution of sodium chloride, a marked increase in p1-f will occur. Ifi cc. of the same solution be added to 10 cc. of a normal solution of sodium acetate, its pH will also increase markedly. In other words the sodium acetate solution is buffered against the addition of an acid, but not against the addition of a base. If, on the other hand, 10 cc. of a normal solution of acetic acid be substituted for the sodium acetate it will be found that this solutionis strongly buffered against the introduction of hydroxyl ions into the solution.

A mixture of equal volumes of molar sodium acetate solution and molar acetic acid solution will exhibit buffer action against both acids and bases over a considerable range of the pH scale.Two important points regarding buffer solutions have been brought out by the foregoing discussion. No one solute as a rule will act as a buffer against both acids and bases. Furthermore no buffer solution will exert a buffer effect over the entire range of the p11 scale.Different buffer solutions vary greatly in their effectiveness in maintaining pH stability. Some are strongly buffered; others weakly. Some are strongly buffered against acids and weakly buffered against bases; of others the converse is true. The commonest types of buffer systems are those composed of a weak acid plus one of its salts. The sodium acetate-acetic acid buffer system already described is such a system. Practically all of the buffer solutions of importance in living organisms belong to this group. Buffer action consists essentially in the tying up of free hydrogen or hydroxyl ions nearly as rapidly as they are introduced into the solution in the formation of compounds which are only slightly dissociated. The ensuing change of pH is therefore relatively small in proportion to the volume of acid or base added. As an illustration let us consider once more a solution consisting of both sodium acetate and acetic acid dissolved in water.

The sodium acetate is strongly dissociated, but the acetic acid will dissociate only slightly, being a weak acid.Suppose now that a little HCl be added to this solution. This is equivalent to adding H_1 and Cl ions and HCl molecules; the latter, however will dissociate, forming additional ions as rapidly as the H_1 and Cl ions already present are bound up in chemical combination. H+ and CH_3COO ions cannot exist side by side in the same solution in appreciable concentrations since CH_3COOH is a poorly dissociated compound. Hence the added H^+ ions are almost all tied up in the formation of CH_3COOH. The Cl ions form NaCl with the Na ions which dissociates in the usual way. The result is that there is only a slight increase in the concentration of hydrogen ions in the solution, and hence only a very slight reduction in pH value. Suppose now that instead of HCl, a little NaOH be added to this solution.This is equivalent to adding Na^+ and OH ions and NaOH molecules; the latter, however, will produce additional ions by dissociating as rapidly as the Na^+ and 0H ions already present are tied up in chemical combination. But OH and H+ ions cannot exist side by side in the same solution in appreciable concentrations since H_20 is only slightly dissociated. Most of the added 0H ions therefore combine with the H1 ions produced by the CH_3COOH and form H_20. More of the CH_3COOH hydrolyzes producing more H+ ions, which in turn unite with more of the OH ions. This continues until practically all of the added OH ions are tied up. The Na1 ions form CH_3COONa with the CH_3COO ions which dissociates in the usual way. The final result is that there is only a very slight decrease in the concentration of H^+ ions in the system, and hence only a very slight increase in its pH value.

Any mechanism which will act in such a way as to remove hydrogen or hydroxyl ions from a solution may operate as a buffer system. Other types of buffering are known but they are relatively of much less importance in living organisms than the type of chemical mechanism which has just been considered. Hydration of Solutes. Molecules of water adhere to the particles of many solutes. Water thus associated with the particles of a solute is called water of hydration. For example, each molecule of dissolved sucrose apparently has six molecules of water associated with it. This means that if a mol of sucrose (342.2 g.) is dissolved in water there will be bound to the sucrose molecules a total of six mols of water (108.096 g.). Ions also are hydrated. Different species of ions appear to have different numbers of water molecules associated with them. Although exact values have been assigned by several experimenters for the number of molecules of water of hydration for certain ions these values are open to question as the results of different investigations do not agree. It seems certain, however, that both anions and cations are hydrated, and that for some kinds of ions, at least, as many as a hundred or more molecules of water of hydration may be associated with a single ion.

Chapter 3

Interfacial Phenomena

There are a number of familiar observations which indicate that the surface layer of water possesses certain distinctive properties which are not exhibited within the body of the liquid. For example, if a perfectly dry, clean, steel needle be carefully laid on the surface of some still water it will float, in spite of the fact that steel is heavier than water. Careful observation shows that the water surface beneath the needle is depressed, forming a tiny liquid cradle, within which the needle is supported. This phenomenon is clearly due to the properties of the surface film of the water, because if the needle is laid on the water at a slight angle, so that this film is punctured, it will immediately sink to the bottom of the vessel. A number of other familiar phenomena are possible principally because of the distinctive properties of the surface film of water. The ability of many kinds of insects to walk on water, the formation of drops by water under certain conditions the rise of water in capillary tubes, soil, blotting paper, etc., the upward bulging of the water surface in a vessel when it is filled to the very last drop, and many other aspects of the behaviour of water, are due largely or entirely to the properties of its surface layer. The surface layers of other liquids also exhibit distinctive properties, but in no common liquid except mercury are such properties as well marked as in water.

SURFACE TENSION

Every molecule within the body of a liquid, although in rapid motion as a result of its kinetic energy attracts and is attracted by other molecules. The attractive forces which any molecule exerts upon its neighbours may be considered as acting along lines of force which radiate from it in all directions. The pull exerted by any molecule upon surrounding molecules, however, diminishes very rapidly with increasing distance from that particular molecule, reaching a maximum at distances not exceeding one or two molecular diameters. Liquids possess a definite volume and distinct boundaries because the molecules are close enough together to attract each other with forces of considerable magnitude. The important property of cohesion in liquids is due to these forces. The magnitude of the cohesive force between the molecules is greater in water than in most liquids. The molecules of a solid are so closely bound together by intermolecular attractions that it has lost ability to change its form, and possesses not only definite, but fixed boundaries. The molecules of both liquids and solids are kinetically active, but their movements are limited more or less rigidly by cohesive forces between the molecules. This limitation upon the freedom of movement of molecules is much greater in solids than in liquids. In gases the distance between the molecules is relatively so great that the intermolecular attractive forces are of negligible magnitude.

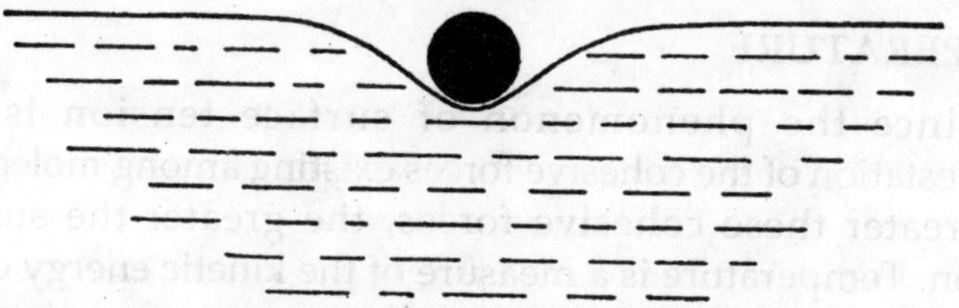

Fig. Needle Floating on water

Molecules at the surface of a liquid or a solid are affected differently bythe cohesive forces between molecules than those within the interior. The further more detailed discussion of this point will be in terms of water. In a vessel of water each of the individual molecules in the body of the liquid is attracted in all directions by the cohesive forces exerted upon it by the surrounding molecules.

At the surface, however, the molecules are not surrounded on all sides by other water molecules. Every molecule in the surface layer is pulled strongly toward the interior by the water molecules beneath it. It is for this reason that drops of water falling freely through space acquire approximately a spherical shape. Since there are no outwardly directed attractions which balance these internally directed forces the surface layer of molecules in any liquid is constantly under a tension. The strained condition of the surface layer of liquid molecules at liquid-gas boundaries is known as surface tension. Solid surfaces also manifest surface tensions, but gases, having enormous intermolecular distances and no surface boundaries, do not. In addition to the unbalanced molecular attractions which the molecules in the surface layer of a liquid sustain, such molecules are usually definitely oriented. This results in a greater concentration of molecules in the surface layer than in the body of the liquid, and contributes toward maintenance of rigidity of the surface film. Surface tension values are generally stated in terms of dynes per centimeter, that is, as the force in dynes which is exerted by the liquid along a line 1 cm. in length.

FACTORS AFFECTING SURFACE TENSION

TEMPERATURE

Since the phenomenon of surface tension is one manifestation of the cohesive forces existing among molecules, the greater these cohesive forces, the greater the surface tension. Temperature is a measure of the kinetic energy of the molecules, hence an increase in temperature means an increase in their kinetic energy and vice versa. As the kinetic activity of water molecules increases,the effectiveness of the cohesive forces existing between them decreases. Hence as the temperature of water increases, its surface tension decreases. The decrease in surface tension of any liquid is, in fact, directly proportional to increase in temperature, except when a liquid approaches its critical temperature. The variation in the surface tension of water within the temperature range at which most physiological processes occur is not sufficient, however to be of any very great significance.

SOLUTES

The surface tension of a liquid is influenced by the presence of the molecules or ions of a solute. Some solute increase the surface tension of water; others decrease it. If alcohol, for example, be added to water the surface tension of the resulting solution will be less than that of pure water. The attraction between an alcohol molecule and a water molecule is less than that between two water molecules. Since the alcohol molecules are distributed among the water molecules, the mean cohesive force in the solution is less than that in pure water. The reduction in surface tension occasioned by the addition of alcohol to the water is one aspect of the reduction of the cohesive forces between the molecules in the solution. Most organic compounds have a similar and often very marked effect on the surface tension of water when dissolved in it. However a few such as sucrose have a slight increasing effect on surface tension. Most inorganic salts when dissolved in water increase its surface tensions. The presence of such solutes increases the cohesion within the liquid by attracting the water molecules more strongly than the water molecules attract each other. The mean cohesive force of the molecules in the solution is thus increased. The result of this effect is a greater surface tension in the solution than in pure water. The magnitude of the effect of inorganic salts upon surface tension is never very great, however.

As a rule, increasing the concentration of a solute in water increases the magnitude of its effect, whether positive or negative, upon surface tension. The influence of solutes upon surface tension is not proportional to their concentration, however, as low concentrations of any solute produce relatively greater effects upon surface tension than high concentrations.

INTERFACIAL TENSION

The condition of tension or strain in a limiting layer of molecules is not confined to boundaries between liquids and gases.At a boundary between two immiscible liquids or between a solid and a liquid each of the abutting layers of molecules is subject to similar tensions. Boun-1 In general compounds which decrease the surface tension of water are of the type known as non-polar, while those which increase it are of the type known as polar.

The term surface tension is usually restricted to tensions developed at the surface of a liquid when in contact with a gas, i.e. at the interface between a liquid and a gas. Surface tension is therefore merely one variety of interfacial tension. The magnitude of interfacial tensions is influenced by the factors of temperature and solutes in the same general way as surface tensions.

ADSORPTION

The molecules of most organic compounds, as we have already seen, have a lesser attraction for water molecules than the molecules of water have for each other. This relatively greater attraction between water molecules than between the molecules of water and those of such non-polar solutes results in the displacement into the surface or interfacial layers of a disproportionately large number of the solute molecules. Hence the concentration of most organic solutes is greater in the surface or interfacial layers than in the body of the liquid. The concentrating of solute molecules in the surface or interface of a liquid is called positive adsorption.

The molecules of inorganic salts, on the other hand, are more strongly attracted to water molecules than the water molecules are attracted to each other. Because of this stronger attraction of the water molecules for such polar solutes relatively more solute molecules are pulled from the surface films into the body of the liquid than water molecules. Molecules of polar solutes, therefore, generally occur in greater concentration in the body of the liquid than in the surface layers. This phenomenon is called negative adsorption. When negative adsorption occurs the solvent molecules become relatively more concentrated in the surface layer than in the body of the liquid.For reasons which will become clear later, negative adsorption rarely occurs at interfaces.The difference in concentration between the surface layer and interior of a liquid are in general much less in negative adsorption than in positive adsorption. Adsorbed molecules are not static, but exhibit a continuous although reduced kinetic activity. Molecules adsorbed at surfaces or interfaces are in dynamic equilibrium with the molecules of the same species in the body of the liquid. An adsorption equilibrium is attained when the number of adsorbed molecules passing out of the interface in a unit of time is equal to the number passing into the interface.

When adsorption occurs from a liquid containing many solutes, as is true of most biological liquids, all solutes are adsorbed to a greater or lesser degree, depending upon their specific properties in relation to the adsorbing surface. In general, however, under such conditions no one solute will be adsorbed as completely as if it alone were present. If a substance is introduced into an aqueous solution which will lower the surface tension of water more than another substance, already adsorbed, the first compound will largely displace the second compound in the interfaces of that solution.

INTERFACIAL ADSORPTION

Adsorption occurs at all types of interfaces.Many solids adsorb gas molecules very powerfully at solid-air interfaces.Solutes may be adsorbed at interfaces between immiscible liquids, and at interfaces between liquids and solids, as well as at liquid-gas interfaces (surfaces). Water molecules are strongly adsorbed at many solid-liquid interfaces. Adsorption is, therefore, a phenomenon of a very general occurrence.

Adsorption at liquid-gas interfaces involves only the forces of attraction between the solvent and solute molecules, since the molecules of a gas are too far apart to have any appreciable attraction for the solute molecules. Similarly adsorption at solid-gas interfaces involves only the forces of attraction between the molecules of the solid and those of the adsorbed gas. At liquid-liquid, or solid-liquid interfaces, however, adsorbed molecules are subject to the attraction of the molecules of the two abutting substances at the interface. At an interface between carbon particles and water, molecules are attracted by both the carbon and the water. If the carbon exerts a greater attraction for the solute than the water as is usually the case a greater accumulation of solute molecules will occur in the interface, than if the opposite is true. The concentration of adsorbed molecules which will be attained at any solid-liquid or liquid-liquid interface is therefore controlled by the relative magnitude of the attractive forces of the two phases for the molecules of the adsorbate (substance adsorbed). Inorganic salts which are negatively adsorbed at surfaces (water-air interfaces) are usually positively adsorbed at water-liquid interfaces, and almost always so at water-solid interfaces.

This is due to a greater attraction for the solute particles by the non-waterphase of the interface than the water molecules themselves exert. Specially treated (activated) charcoal is one of the best adsorbents known both for gases and for solutes. Many important applications are made of this property of charcoal in chemistry and technology. Gas masks for protection against poison gases in warfare owe their efficiency to the adsorptive capacity of charcoal. Colored or other soluble impurities are often removed from liquids by passing them through layers of charcoal. Solids such as charcoal which are powerful adsorbents usually possess an enormous surface area in proportion to their mass, which accounts for their effectiveness as adsorbing agents. The surface of g. of charcoal has been variously estimated at from about 50 to about 600 square meters. Much of this surface is in the form of internal submicroscopic.

Adsorption in which the fundamental controlling forces are the cohesive and adhesive forces between molecules is called mechanical adsorption in order to distinguish it from certain more complex kinds of adsorption phenomena. Of these the most important is electrical adsorption. Most surfaces in contact with water bear an electrical charge. The presence of such charges markedly influences the adsorption of ions or other charged particles. Cellulose, in common with many other substances, acquires a negative charge when immersed in water. If a strip of cellulose filter paper be arranged so that its lower end dips in a solution of eosin (a red dye) it will be observed that the eosin will rise through the filter paper almost as rapidly as the water from the solution rises by capillarity. If another strip of filter paper be similarly arranged to dip in a solution of methylene blue (a blue dye) the water will rise up the filter paper as rapidly as it does from the eosin solution, but the dye will scarcely rise at all. The difference in behaviour of these two dyes is due principally to the electrical charge which their coloured ions carry. The coloured ions of eosin are negatively charged; those of methylene blue positively charged. The negatively charged particles of eosin are repelled from the negatively charged surface of the cellulose fibers. The eosin particles are therefore driven towards the centre of the capillary columns of water in between the fibers, and move up the filter paper almost as rapidly as the water. As soon as the positively charged particles of methylene blue come in contact with the negative charges of the cellulose, however, they are held to the surface of the cellulose by the forces of electrical attraction. In other words they are electrically adsorbed. While the water rises by capillarity at about the same rate as it does from the eosin solution, the rise of methylene blue is relatively very slow.

Certain examples of adsorption have also been recognized which are designated by the name of chemical adsorption. Chemical reactions are involved in adsorption phenomena of this type. The familiar reaction of 12K1 with starch in which the starch stains purple is often considered to be an example of chemical adsorption. Actually there exists no clean cut distinction between mechanical, electrical, and chemical adsorption. The several phenomena intergrade, and elements of more than one of these mechanisms are usually operating whenever adsorption occurs. Since the surface or interface of even pure water bears an electric charge, at least minor electrical effects are involved in even the simplest adsorption phenomena.

ADSORPTION OF WATER

Water itself is strongly adsorbed at certain types of interfaces. The adsorption of water is a phenomenon which can be regarded as more nearly similar to electrical adsorption than to mechanical adsorption. Water molecules, while not charged in the sense that ions are strongly polar and behave somewhat as charged bodies. Water-vapour molecules become adsorbed on many different kinds of surfaces. The water-vapour adsorbed upon glass weighing bottles often becomes an appreciable source of error in accurate quantitative work. The particles in certain colloidal systems have the property of adsorbing large quantities ofwater, as shown in the next two chapters. A large portion of the water which moves into various types of substances in the process of imbibition becomes adsorbed upon the internal surfaces of the imbibing substance.

THE BIOLOGICAL SIGNIFICANCE OF ADSORPTION

Adsorption phenomena are known to play a manifold role in living organisms, probably being involved in practically all cell activities. Protoplasm and many other constituents of plant cells are essentially colloidal, and adsorption phenomena are of general occurrence in colloidal systems. Within plant cells many interfaces occur, as at the boundaries between the protoplasm and vacuole, protoplasm and cell wall, and nucleus and cytoplasm, at all of which interfacial concentration of solutes undoubtedly occurs. The adsorption of certain compounds at the cytoplasmic interfaces is generally believed to exert a marked influence on the permeability of the cytoplasm. It is quite possible that interfacial tensions play an important role in the process of cell division. Imbibitional phenomena, of basic importance in the water relations of plant cells, involve the adsorption of water. The action of enzymes as well as of other catalysts is generally believed to involve adsorption phenomena. Much information regarding the structure of cells has been gained by the use of dyes which are differentially adsorbed by various constituents of cells. Chromosomes, for example, are so named because they strongly adsorb certain stains.

COLLATERAL READING

The relation of adsorption to many of the processes and phenomena mentioned above will receive further attention in subsequent chapters of this book. There are probably few if any processes occurring in living organisms in which adsorption phenomena are not at least indirectly involved. Adsorption and other interfacial phenomena are responsible for many of the so-called vital activities of living cells.

Chapter 4

Colloidal Systems

Most biological problems lead ultimately to a consideration of one phase or another of the more fundamental problems of the structure, constitution, and physico-chemical properties of protoplasm. Certain characteristic physicochemical properties of protoplasm have been recognized and studied. Protoplasm is more or less elastic. It may vary in viscosity from values not much greater than that of pure water to those of jellies. It possesses a pronounced capacity for the imbibition of water. Protoplasm is usually high in water content, but nevertheless often behaves as if immiscible with water. Mechanical disturbances may markedly influence its physical state. When exposed to very high or very low temperatures, to high concentrations of salts or to certain other factors, protoplasm may be irreversibly coagulated. Under certain conditions it may lose its fluidity and assume a jelly-like condition, a change in physical state which is usually reversible.

Protoplasm is predominantly composed of substances in the colloidal state and it is to these colloidal systems that it owes most of its characteristic physico-chemical properties, some of the better known of which have been listed above. The world of living organisms has, in fact, been largely molded on a colloidal pattern. Most physiological processes, reduced to their ultimate tangible mechanism, either occur in a colloidal matrix, or under such conditions as to be strongly influenced by the colloidal organization of the cells in which they take place. Many physiological processes occur only under the influence of the organic catalysts known as enzymes, which are supposed by most authorities to be in the colloidal condition and to owe many of their properties to that fact. It is impossible, therefore, to obtain any adequate comprehension of the properties of protoplasm without a background of facts and principles regarding colloidal systems. The more fundamental properties of such systems will be discussed in this and the following chapter. Some of the most important components of the material environment of plants and animals are also essentially colloidal. Most soils contain a considerable proportion of matter in the colloidal or near-colloidal condition, and owe many of their most distinctive properties to this fact. Few streams or bodies of water are entirely free from matter in the colloidal condition. Clouds, fogs, mists, and smoke also represent matter in the colloidal state.

GENERAL NATURE OF COLLOIDAL SYSTEMS

If a little sucrose or sodium chloride be shaken up in water the solid will soon disappear, and the resulting system will be a solution. Although there is abundant evidence which indicates that the solute particles are dispersed throughout the solvent, molecules and ions are so small that it is impossible to detect their presence by direct observation, even with the most powerful optical systems available. If, instead of sucrose or sodium chloride, we stir some line river bottomsilt into water, a different sort of a system will result. The silt particles don't dissolve but simply become dispersed throughout the liquid. Such a system is called a suspension. The particles in it are of such size that they can readily be detected under a microscope. Suspensions are not stable, however, as the particles gradually settle out, and the system becomes separated into its two original components within a relatively short period of time. Similar systems can be prepared by vigorously shaking together two immiscible liquids such as oil and water. Such systems are called emulsions. Emulsions are not stable unless there is also present in the system a third component called an emulsifier.

Still another type of system consisting of particles dispersed through water can be prepared. If a trace of sulphur be dissolved in a small volume of alcohol which is then poured into a some what larger volume of water, a cloudy opalescent liquid will result which is composed of sulphur particles dispersed through water. This type of system is intermediate in its properties between solutions and suspensions. The dispersed particles are not molecules, but, as in suspensions, aggregates of molecules. Unlike suspensions, however, such systems are relatively stable, as the particles will remain dispersed throughout the liquid indefinitely. The dispersed particles of such systems cannot be seen under a microscope, showing that they are smaller than the particles in suspensions or emulsions. They can, however, be detected with the aid of an ultramicroscope. The sulphur-in-water system which has just been described is a simple example of a colloidal system.

Colloidal systems, as the preceding discussion has indicated, are two phased systems like solutions. Unlike solutions, however, the particles of the dispersed phase are not in the molecular or ionic condition, but with certain exceptions to be noted shortly are molecular aggregates. One colloidal particle is often composed of hundreds or even thousands of molecules lumped together. The molecular aggregates must not be so large however, that the particles readily settle out of the system, as stability is one of the essential attributes of colloidal systems.

Hence molecular dispersions of such substances are simultaneously both solutions and colloidal systems. The limits generally accepted for the range of sizes of colloidal particles have been somewhat arbitrarily set and actually there is no sharp boundary between colloidal systems and suspensions on the one hand, nor between colloidal systems and solutions on the other.There is a perfect gradation in properties from one type of system to the next. The properties of suspensions or emulsions in which the suspended particles are of small dimensions approach those of colloidal systems, while the smaller the dispersed particles in a colloidal system the more closely it approaches a solution in its properties.

The term to designate a certain group of substances, which seemed to be set apart from other substances by several distinctive properties. When dispersed in water these substances had a slow rate of diffusion, and failed to diffuse through membranes of parchment paper. Furthermore they did not form crystals. He applied the alternate term crystalloid to those crystal-forming substances which, when in solution, diffused relatively rapidly and passed readily through parchment membranes. We now know that, strictly speaking,no such distinction can be made; the word colloid properly refers to a distinctive state of matter, and cannot be applied with accuracy to any one class of substances. Theoretically any substance can, by proper manipulation, be brought into the colloidal state, and actually this experience fall naturally into two groups generally known as oil-in-water emulsions, and water-in-oil emulsions. In the former type an oil or some other liquid insoluble in water, or practically so, is dispersed throughout a water dispersion medium. In the latter type the converse is true, the oil or other liquid immiscible with water constitutes the dispersion medium, while small aggregations of water molecules are dispersed through it. The proportions of the components of most emulsions can be varied between wide limits.Emulsions are not generally considered to be true colloidal systems, but, like suspensions, approach them in properties. Emulsions occur commonly in the cells of plants and animals, and are generally believed to be essential components of the protoplasmic matrix. Both water-in-oil and oil-in-water emulsions are known to exist in living cells, but the latter type is commoner. When examined under high magnification with a microscope protoplasm in its grosser aspects often presents the appearance of an emulsion of fats and fat-like substances dispersed through the body of the protoplasm.

Some common examples of oil-in-water emulsions are milk, cream, emulsions of olive oil in water, and the latex of the rubber tree, milk weed, etc. Perhaps the only generally familiar water-in-oil emulsion is butter. Many other similar systems can be prepared, however. Water in olive oil is one example of such a system. Water in kerosene is another. The word oil as employed in this discussion is not restricted to those substances usually classified chemically as oils, but is here somewhat generalized to include other liquids which are immiscible with water. Kerosene, for example, is not chemically an oil.

Emulsions with the exception of some very dilute ones, lack stability unless there is also present in the system an emulsifier. In the absence of an emulsifier, the two components of an emulsion rapidly separate, the oil, being the component of lower specific gravity rising to the top. The group of substances classified as emulsiflers is chemically a very heterogeneous one. Some of the best known emulsifiers are the soaps, saponins, various substances which form hydrophilic colloids when dispersed in water (gums, gelatin, etc.), andfine suspensions of certain rather inert materials such as sulphur, carbon, silica and resin. The last group is probably of little importance in living organisms.Oil-in-water emulsions may be stabilized by such substances as soaps of the alkali metals (Na, K, Li, etc.), gum acacia, and proteins; while soaps ofthe alkaline earth metals (Ca, Ba, etc.), and gum dammar are examples of stabilizers of water-in-oil emulsions. In general, emulsifying agents which are soluble in water or form hydrophilic sols in water, stabilize oil-in-water emulsions while those which are insoluble in water but soluble in oil stabilize water-in-oil emulsions.Many pharmaceutical preparations are stabilized by means of gum acacia or gum dammar.

Emulsions found in nature are usually stabilized by proteins. This is true of milk and cream, in which the droplets of fat are dispersed throughout a water medium. The latex of plants, one of the best examples of a naturally occurring emulsion, is also stabilized by proteins. Mayonnaise dressing is essentially an emulsion of olive oil in water, stabilized by the proteins of the egg which is the third essential component of the system. Protoplasmic emulsions are probably also stabilized by proteins, although soaps may also act a emulsifiers in plant cells.

The mechanism of the stabilizing effect of emulsifiers upon emulsions is not well understood, and it is probable that it varies with different types of emulsifiers. It is well known, however, that any substance which greatly lowers the interfacial tension between water and another liquid will usually stabilize an emulsion composed of those two liquids. The usual interfacial tension between water and benzene, to cite one example, is about 35 dynes per centimeter. Addition of a little sodium oleate (a soap) to water will lower this interfacial tension to about 2 dynes per centimeter. Sodium oleate is an excellent stabilizing agent for an emulsion of benzene in water.

Chapter 5

The Properties of Sols and Gels

The discussion in this chapter will deal only with sols in which water is the dispersion medium. Such systems are classified into the two groups of hydrophilic sols and hydrophobic sols. The dispersed particles in all such colloidal systems fall within the size range of 0.001 to 0.1, but their size may vary greatly with different sols. Even in a given sol the particles may exhibit a wide range of dimensions, although in some they are quite homogeneous in size. Most of the important differences between hydrophilic and hydrophobic sols are due to the hydration of the dispersed particles in systems of the former type. There is no general agreement regarding the exact physico-chemical relationship between the micelle and its water of hydration; but only two conceptions have any wide currency at the present time.

It is presumed that the first layer of oriented water molecules is so firmly attached to the micelle that it is essentially an integral part of it. The successively enveloping layers of water molecules are also oriented, but with increasing distance from the periphery of the micelle the forces of attraction and orientation progressively decrease. In this zone there is a gradual transition from water molecules which are practically all oriented to those which are completely unoriented. Oriented molecules fit together more closely than unoriented molecules. They are packed more closely, just as more bricks can be stacked in a given space if arranged regularly, than if tossed in indiscriminately. As a result of this packing the water in these oriented shells has a greater density than that in the bulk of the liquid;in other words there is an actual contraction in the volume of the liquid associated with the micelles.

The fact that micelles may, under certain conditions, lose their water of hydration very rapidly would seem to favour the latter theory. It is possible of course, that in some hydrophilic systems the water is actually dissolved in the micelles, that in others it is present only as a shell of oriented molecules while in still others both of these two suggested modes of hydration may exist.The important properties of sols will now be summarized, with special attention to differences between the properties of sols of the hydrophilic type and sols of the hydrophobic type.

DILUTION

Most sols are extremely dilute. In other words the actual mass of substance dispersed throughout the dispersion medium is extremely small in proportion to the total volume of the system. The well-known colloidal gold sols, for example, rarely contain more than i g. of gold dispersed per liter of sol. Other hydrophobic sols are correspondingly dilute.

Most hydrophilic sols are also extremely dilute, although there are some exceptions to this statement, as very viscous sols of such substances as starch and gum acacia can be prepared which contain parts or more of dispersed material per 100 parts of sol.

SLOW RATE OF DIFFUSION OF THE DISPERSED PARTICLES

In general the particles which make up the disperse phase of a sol diffuse much more slowly than substances in true solution, i.e. in the molecular or ionic state. The slow rate of diffusion of micelles as compared with most ions and molecules is clearly correlated with the relatively large size of the colloidal particles.There is, however, no sharp line of demarcation in terms of diffusion rates between colloidal micelles and solutes. The diffusion rates of some solutes with large molecules are not perceptibly faster than those of the dispersed particles of colloidal systems in which the micelles are relatively small in size.

FILTERABILITY

Sols are usually filterable; that is they pass through ordinary filter papers without any appreciable separation of the disperse phase from the dispersion medium by the filter. Usually there is some loss of the disperse phase due to an initial adsorption when the sol first comes in contact with the filter. Sometimes this may be very considerable. Since the pores in ordinary filter papers are about 2–5 L in diameter, and even porcelain filters such as those widely used in bacteriological work, have pores 0.2–0.6 in diameter, it is easy to understand why micelles with diameters in the size range 0.001–0.1 pass through.

Special filters have been devised, however, with pores of such a small diameter that the disperse phase can be separated from the dispersion medium by filtering a sol through them. Such filters are known as ultra filters. The process of filtering through such a filter is known as ultrafiltration. The most commonly used types of ultrafilters are those made of collodion or gelatin.The size of the pores in such filters can be controlled by the length of time allowed for drying and in other ways. It is possible to prepare ultra filters with pores of such dimensions that solutes can pass through them, but colloidal micelles cannot.

TYNDALL PHENOMENON

Suppose that a clean glass vessel, preferably one with flat, parallel sides, be filled with pure water and held so that a strong pencil of light passes laterally through the vessel. If an observation be made laterally and at right angles to the path of the beam of light, no distinctive trace of its path through the water can be detected. Such a liquid is said to be optically empty. The same will be true if the water in the vessel be replaced by a sugar or salt solution, or in fact, by any true solution. Suppose however, that the vessel be filled with a hydrophobic sol and observed in the manner just described. The results are strikingly different. The path of the light will be clearly delineated as a cloudy, often opalescent track through the sol. Even a colloidal system which seems perfectly transparent to the unaided eye will usually show some turbidity when submitted to this test. The intensity of this effect varies greatly with the specific colloidal system, and with the concentration of the disperse phase, but it is universally shown by hydrophobic sols.The phenomenon just described is called the Tyndall phenomenon, and is due to the scattering or diffraction of light. The difference in the index of refraction between the two phases of the colloidal system is also a factor determining the intensity of the Tyndall effect. The greater this difference the stronger the effect. Since in the diffraction of light, the short wave lengths (blue end of the spectrum) are bent more than the longer wave lengths, a partial separation of the spectrum results., For this reason a sol with a colourless disperse phase often appears to be pale blue when viewed in the path of a strong beam of light.

Similar Tyndall phenomena are exhibited by hydrophilic sols, but usually the effect is less striking when sols of this type are viewed in the path of a beam of light than the effect observed when hydrophobic sols are employed. Actually a trace of the light track will usually be perceived even in water or true solutions, due to the presence of contaminating dust particles. In order to prepare a truly optically empty liquid, provision must be made for the removal of such dust particles.

The instrument known as the ultramicroscope is based on the principle of the Tyndall phenomenon. It is not possible to observe colloidal particles directly in the ultramicroscope; only the light diffracted from their surfaces can be seen. Neither can any definite image of particles in this small range of sizes be obtained.The ultramicroscope is a microscope which is so arranged that the colloidal system or other material to be examined can be illuminated laterally (i.e. at right angles to the tube of the microscope). This lateral illumination is usually provided by a powerful source of light and a suitable series of condensing and focussing lenses, so arranged that the light is focussed to a point within the mount. Under the ultramicroscope the dispersed particles of a hydrophobic sol appear as bright spots of light varying in size and brilliancy. Very little concerning the actual size or shape of the micelles can be determined since each bright spot represents merely the light diffracted by a single particle. It is possible, however, to determine the number of particles in a given volume of solution by means of the ultramicroscope. The use of the ultramicroscope consists essentially in an observation of the Tyndall phenomenon in a small volume of a sol under the microscope.

BROWNIAN MOVEMENT

In 1828, the botanist Robert Brown observed through a microscope that pollen grains which were suspended in water showed a rapid oscillatory motion. Brown at first was inclined to attribute this motion to the fact that the pollen grains were alive, but examination of preparations of dead pollen grains and spores showed that they likewise exhibited such a motion. It became evident therefore that this movement was in no way connected with living processes. This phenomenon is termed Brownian movement, after its discoverer. Many suspensions in which the particles are within the range of microscopic visibility exhibit Brownian movement. It is clearly shown by many of the smaller species of bacteria when suspended in water. In solid-in-gas colloids, such as tobacco smoke, the dispersed particles show a very vigourous Brownian movement. Particles in the protoplasm of slime molds and certain other species frequently exhibit a Brownian movement which is clearly discernible under the microscope. For particles of a given mass, the smaller their volume the greater the amplitude of their Brownian movement. For particles of equal volume, the less their mass the more vigorously they will exhibit Brownian movement. In general this phenomenon is exhibited more clearly by the micelles of hydrophobic sols than by those of hydrophilic sols.

The viscosity of the liquid phase is also an important factor governing the rapidity with which the dispersed particles move. The more viscous the liquid, the more sluggish the movement of the particles.Brownian movement is caused by the kinetic activity of the molecules of the solvent. Even the smallest particles in which Brownian movement can be observed are very large in proportion to the size of the solvent molecules which impinge upon them. A particle suspended in a liquid such as water suffers a continual bombardment by the molecules of the liquid. If the particle be relatively large, at any given moment it is bombarded on every side by numerous molecules, moving in all possible directions, and at different speeds. The effects of the individual impacts largely counteract each other however, and there is little or no movement of the particle. If the particle be smaller, however, the results are quite different. At any given moment a much smaller number of water molecules impinge upon the particle. The resulting forces cease to be balanced and the sum total effect of the blows which the particle sustains on some one side are greater than the effect of the blows sustained on any other side. Hence the particle moves. The next moment a greater impetus may be given to the particle from some other direction and the course of its movement is changed. In this way the highly erratic movements of suspended particles known as Brownian movement originate. Increase in temperature increases the rate of Brownian movement because of an increase in the kinetic energy of the solvent molecules. This phenomenon is the nearest approach we have to actual visible evidence of the validity of the kinetic theory of matter. It almost brings before our eyes the veritable dance of the molecules.

OSMOTIC PRESSURE

The osmotic pressure of solutions is a colligative property; that is one which depends on the proportion of solute particles (molecules or ions or both) to solvent molecules regardless of the kind of solute particles. A sol should also possess an osmotic pressure since theoretically a micelle has the same effect upon the magnitude of the osmotic pressure as a molecule or an ion. A brief discussion should aid in making this point clear. Large molecules have the same theoretical effect upon osmotic pressure as small molecules. The individual molecules of some compounds are large enough to fall within the colloidal range of sizes. They are both molecules and micelles at one and the same time. Such a micelle should have just as much of an effect on osmotic pressure as a molecule. Actually the osmotic pressures of sols never exceed a small fraction of an atmosphere and are therefore much less than the osmotic pressure of any but the most extremely dilute solutions. The osmotic pressure of colloidal systems depends upon the concentration of the particles of the disperse phase. In most sols the total mass of the disperse phase is not only relatively small; but it is present in the form of particles which are much larger than molecules or ions. The negligible osmotic pressures of sols area necessary corollary of the relatively small concentration of dispersed particles.

VISCOSITY

The viscosity of a fluid is its resistance to flow. The more viscous a liquid the less readily it will flow. Glycerine, for example, is much more viscous than water. The viscosity of hydrophobic sols never varies appreciably from that of the dispersion medium water. Unlike hydrophobic sols the viscosity of hydrophilic sols is usually greater than that of the dispersion medium. The viscosity of hydrophilic sols increases appreciably with increase in the concentration of the sol, but the relation is not a linear one to the concentration of the viscosity of all liquids, including sols, is disperse phase.influenced by temperature. In general, increase in temperature decreases viscosity. In hydrophilic systems this reduction in viscosity with increase in temperature is probably due to two factors: the decrease in viscosity of the medium itself, and the decrease in the hydration of the micelles.

ELECTRICAL PROPERTIES

The dispersed particles of all hydrophobic sols carry electrical charges. A colloidal system, however, is electrically neutral because for every charge carried on a micelle an equal charge of opposite value is carried by ions in the dispersion medium. The situation is similar to that in a solution of an electrolyte. Although the individual ions are charged for every negative charge carried by an anion an equal positive charge is carried by a cation. In some colloidal systems the dispersed particles are negatively charged, in others positively charged, but normally all the dispersed particles in any one system bear a charge of the same sign.

Whatever the origin of the micellar charges they invariably are produced in such a way as to involve the release of ions into the dispersion medium which may therefore be regarded as also being charged. When the micelles are negatively charged, the dispersion medium is positively charged and vice versa. Electrostatic attraction therefore exists between the surface charges of a colloidal particle and the ions of opposite charge in the dispersion medium.The result is that surrounding each colloidal particle with its charged surface is a shell of ions of opposite charge.

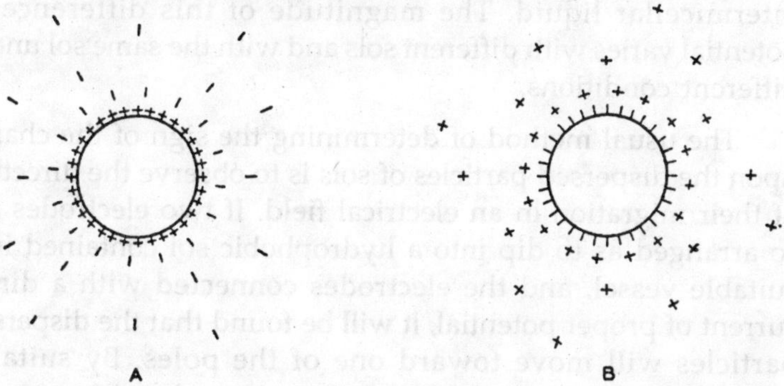

Fig. Diagrammatic representation of the electric charge around micelles: (A) positively charged, (B) negative charged

This arrangement of charges at the surface of a micelle is called an electric double layer. Similar electric double layers exist also at boundaries between solid surfaces and liquids, as for example along the walls of a capillary tube. One positively charged, one negatively charged. The innermost layer of ions in the dispersion medium is probably compactly oriented around the oppositely charged micelle, which is in turn surrounded by progressively more diffuse layers. That is, while most of the ions are close to the charged surface of the micelle, some are farther away, although with increasing distance from the surface of a micelle, the number of ions associated with that micelle decreases rapidly.

The ions of the double layer are in dynamic equilibrium with undissociated molecules at the periphery of the micelle proper. Anions and cations of the double layer are continually uniting and forming uncharged molecules which become part of the micelle. Contrariwise molecules at the surface of the micelle are continually dissociating into cations and anions. If the micelle is negatively charged the anions adhere to its surface; if positively charged, the cations. The ions of the opposite charge become part of the outer shell around the micelle.

The presence of this electrical double layer results in a difference of electrical potential between the micelle and the intermicellar liquid. The magnitude of this difference of potential varies with different sols and with the same sol under different conditions.

The usual method of determining the sign of the charge upon the dispersed particles of sols is to observe the direction of their migration in an electrical field. If two electrodes are so arranged as to dip into a hydrophobic sol contained in a suitable vessel, and the electrodes connected with a direct current of proper potential, it will be found that the dispersed particles will move toward one of the poles. By suitable arrangements, using a microscope or ultramicroscope (depending upon the size of the particles), the migration of the particles may be actually watched. In a positively charged ferric hydroxide sol, for example, the micelles will move towards the negatively charged electrode, while in a negatively charged arsenious sulfide sol they will move towards the positively charged pole. This phenomenon is known as cataphoresis or electrophoresis.

As a micelle moves under the influence of an electric current only the Innermost layer of the electric double layer the one which determines its electrical charge clings to the micelle and flow with it. This inner layerslides past the oppositely charged ions of the outer shell of one micelle after another as it moves towards the anode if negatively charged, or towards the cathode if positively charged. Simultaneously ions of the outer layer move towards the opposite pole from the one towards which the micelles migrate.There is a close analogy between this phenomenon and the behaviour of the ions of an electrolyte during electrolysis.

The sign of the charge on the dispersed particles of any sol can be determined by cataphoresis. Particles considerably above the colloidal range of dimensions frequently can be shown to exhibit cataphoretic migration. The phenomenon can be demonstrated with bacteria, unicellular algae, spores, etc.Practically all such small living organisms are negatively charged.

The dispersed particles of hydrophobic sols apparently acquire their charges either by electrolytic dissociation or by adsorption. In some such systems the charges seem to arise as a result of ionization of some of the molecules composing the micelle. The ions released into the dispersion medium become the outer envelope of the double layer, leaving the micelle with a residual and compensating charge of ions of opposite sign. Individual molecules of some substances are large enough to fall within the colloidal range of sizes. The dye Congo Red, which is the sodium salt of a complex organic acid, is an example of such a substance. Dispersed in water this compound produces sodium ions and a colloidal anion, the latter, of course, being negatively charged.

In other systems the electrical charge on the dispersed particles is apparently acquired by adsorption of either the positive or negative ions of an electrolyte, the ion of opposite charge becoming the outer shell of the double layer. Ferric hydroxide sols 2 are normally positively charged. The charge on the micelles of these sols is often ascribed to the adsorption of+ 1 ions of the $FeCl_3$ from which ferric hydroxide sols are usually prepared, the Cl ions forming the outer layer around the micelles. Similarly the negative charge of the micelles of arsenious sulfide sols is often ascribed to the adsorption of H_25 which is used in the preparation of this sol. Dissociation of H_25 results in the release of H^+ ions into the dispersion medium leaving the dispersed particles with a residual and compensating negative charge.

Similarly it is believed that some substances acquire an electrical charge by adsorbing hydrogen or hydroxyl ions more frequently the latter from their water dispersion medium. Certain inert substances such as cellulose,carbon, quartz, collodion, etc., are believed to become charged in this way.All of these substances acquire a negative charge when in contact with water,indicating that the hydroxyl ions are adsorbed, the hydrogen ions becoming the outer shell of the double layer.

The micelles of some Izydrophilic sols are charged; those of others are not.As in hydrophobic systems the dispersed particles of different hydrophilic sols acquire their charges in different ways. In some the charges on the particles 2 Most investigators believe that the so-called ferric hydroxide sol is actually a sol of hydrated ferric oxide ($Fe_2O_3(H_2O)$).

The electrical charges on the micelles of protein sols arise in this way. In othersystems the charges may originate from traces of electrolytes which are present as impurities. This is probably true of agar sols. Such charges may be regarded as similar in their origin to those acquired by adsorption on the micelles of hydrophobic sols, in that the charge is due to some compound associated with the substance of which the micelle is composed, and not to that substance itself. Most of the better known hydrophilic sols are negatively charged.

FLOCCULATION

Since the dispersed particles of any sol are in rapid motion it would seem that repeated collisions would result in a progressive agglomeration of the particles into larger and larger masses which eventually would settle out of the system. Sols, however, are relatively stable systems, and it is important to understand the mechanism by which the stability of such colloidal systems is maintained.The stability of hydrophobic sols is maintained almost entirely by the charge which each micelle carries. Although by Brownian movement the dispersed particles are repeatedly brought close together actual collision seldom occurs because the shells of ions around the micelles are mutually repellent.

Reduction of the charge on the micelles of any hydrophobic sol to the point where there is no difference of electrical potential across the double layer results in agglomeration of the individual particles into flakes of a size which rapidly settle out of the surrounding liquid. This phenomenon is called flocculation, coagulation, or precipitation. The point at which there is no difference of electrical potential across the double layer is known as the iso electric point of the sol. At the isoelectric point the micelles of a sol are,relative to the surrounding medium, completely uncharged. As, by Brownian movement, two such micelles are brought into contiguity they no longer repel each other with sufficient intensity to prevent their agglomeration. By the addition of other micelles such particles rapidly increase in size, soon resulting in flocculation of the sol. Merely reducing the electric charge to a value approaching that of the isoelectric point is sufficient to induce instability and slow flocculation in many sols.Flocculation is most commonly initiated by the introduction of electrolytes into the system. Very small quantities of an electrolyte are often sufficient to cause the flocculation of a relatively large volume of sol. The important principles regarding the flocculation of hydrophobic sols by electrolytes canbe most easily elucidated by a discussion of some of the data in. The flocculating effect of an electrolyte is due primarily to the added ion of opposite charge from that borne by the colloidal particle. Thus the As_2S_3sol may be flocculated by cations such as Na1, Ca^{+1}, or Al^{+11}, while theFe(OH)3 sol may be flocculated by anions such as C1, NO_3, or SO_4^-.

Furthermore, the flocculating effect increases with an increase in the valency of the effective ion.

However, the influence of the valency of an ion upon its flocculating effectiveness does not follow a simple 1 2 3 arithmetical ratio. In the flocculation of hydrophobic sols the ion of the added electrolyte with a charge of the same sign as that of the micelle is not entirely without effect. The influence of such ions is usually to increase the stability of the system. Precisely stated, therefore, the influence of an electrolyte upon the micelles of a hydrophobic sol is a differential effect between the anions and cations but the influence of the ion of opposite charge predominates.The mechanism of the flocculation of a hydrophobic sol is too complex a process to be considered in more than a general way in an introductory discussion. In general flocculation results from destruction of the double layers around the micelles. There is a close analogy between the flocculation of a hydrophobic sol by an electrolyte and a precipitation reaction between one electrolyte and another. The micelles of hydrophobic sols may be regarded as giant ions bearing numerous charges.

When $CaCl_2$ is added to a solution of Na_2SO_4 slightly dissociated $CaSO_4$ is precipitated, leaving paired Na1 and Cl ions in the solution. An analogous phenomenon results when $CaCl_2$ is added to a negatively charged As_2S_3 sol in which the outer shells of the double layers are composed of hydrogen ions derived from the dissociation of adsorbed H_25. The resulting flocculant consists of uncharged particles of As_2S_3 + Ca, while the H^+ and Cl ions left behind in the solution pair off in the usual manner, forming hydrochloric acid. In other words Ca^{++} ions have replaced the H + ions in the outer shells of the micelles. The result, however, is an unstable particle, since Ca1 1 ions do not remain in the dissociated state when associated with As_2S_3 micelles which owe their negative charge to the dissociaion of adsorbed H_2S molecules. They combine with such micelles, thus dissipating their electrical charges. Flocculation of the uncharged micelles then ensues.

Flocculation may also be initiated by introducing into a hydrophobic sol another hydrophobic sol with micelles bearing a charge of opposite sign. If $Fe(OH)_3$ sol be slowly added to an As_2S_3 sol a point will be reached at which complete flocculation will occur. The particles of the two sols will settle out as an intimate mixture. The same phenomenon occurs when any negatively charged hydrophobic sol is added to any positively charged hydrophobic sol in sufficient quantity, or vice versa. This process is called mutual flocculation. When this phenomenon occurs the ions of the outer zone of one kind of particle pair off with the oppositely charged ions of the outer shell of the other kind.

The addition of a small amount of a hydrophilic sol, such as a gelatin or gum arabic sol, to a hydrophobic sol makes flocculation of the latter by electrolytes or micelles of opposite charge difficult or impossible. This effect of a hydrophilic on a hydrophobic sol is termed protective action. Protective action is apparently due to the adsorption of the micelles of the hydrophilic sol around the micelles of the hydrophobic sol. The properties of the sol,therefore, become essentially those of the hydrophilic system. As will be seen shortly hydrophilic sols are much less easily flocculated by electrolytes than hydrophobic sols and this is undoubtedly the basis for protective action.We turn now to the question of the mechanism of the flocculation of hydrophilic sols. The micelles of such sols may or may not carry an electrical charge, but whether charged or not such colloidal systems are stable. Hydrophobic sols, as already shown, are stable only when the micelles bear an electrical charge. One of the most important differences between hydrophobic and hydrophilic sols is the possession by the latter of a second stability factor,which in itself is effective in keeping such sols stable for long periods of time.In our previous discussion we have seen that uncharged micelles of hydrophobic sols soon agglomerate and settle out of the dispersion medium. Why do not the uncharged micelles of a hydrophilic sol behave in the same way?This is probably due to the effect of hydration upon the properties of the dispersed particles. The micelles of all hydrophilic sols, it will be remembered, are highly hydrated. The first layer of molecules of water of hydration is probably so firmly bound to the particle as to virtually constitute an integral part of the micelle itself. Surrounding this are other layers of water molecules more or less completely oriented depending on their distance from the surface of the particles. Each such particle is cushioned against impacts with other particles by its enveloping shell of oriented water molecules.Agglomeration of the micelles is thus prevented, and hence even uncharged hydrophilic sols are stable.

The possession of two stability factors by hydrophilic sols complicates the mechanism of flocculation in such systems. The manner in which flocculation of hydrophilic sols may occur can be illustrated by reviewing a simple experiment. The experimental material is a dilute (about o.1 per cent) sol of agar-agar. Such a sol is perhaps the most typical example of a simple hydrophilic system. If a relatively large volume of alcohol be added to a small portion of such a sol, the micelles lose their water of hydration, and the sol acquires the cloudy, bluish, opalescent appearance typical of many lyophobic sols. In fact it now is a lyophobic sol, and shows all the typical properties of such systems. The alcohol, which is a powerful dehydrating agent, has robbed the micelles of their shells of oriented water molecules. Nevertheless, the sol is still stable, since the micelles retain their original negative charges. Finally, let a drop of a solution of an electrolyte such as $AlCl_3$ be added to the sol. The sol now flocculates almost immediately since the only remaining stability factor the electrical charge has been destroyed by the addition of the electrolyte, and the opalescent cast of the sol disappears.The stability factors of a hydrophilic sol can also be dissipated in the opposite order. Suppose that the $AlCl_3$ solution first be added to the agar sol until the charges on the micelles are neutralized. Although now at its isoelectric point, unlike hydrophobic systems, the sol does not flocculate. If however, alcohol now be added to the system, immediate flocculation occurs because of a dehydration of the micelles, resulting in an elimination of the only remaining stability factor in the system.

Briefly then, in order to flocculate a hydrophilic sol, its micelles must be both dehydrated and electrically discharged, except in the occasional systems in which micelles are uncharged, in which dehydration alone will suffice. Dehydration alone of a hydrophilic sol with charged micelles results in its conversion into a lyophobic sol.

The addition of relatively large quantities of certain electrolytes to hydrophilic sols will result in their flocculation without any previous removal of the water of hydration of the micelles by means of alcohol or any other dehydrating agent. Only very soluble salts are effective in this way. It is evident that this phenomenon, usually called salting out, involves a more complicated reaction than that taking place in simple flocculation and must be distinguished from the latter phenomenon. Three salts which are especially suitable for salting out hydrophilic sols are $(NH_4)_2S0_4$, $MgSO_4$, and Na_2SO_4. In these three salts the ions, especially the anions, acquire relatively large quantities of water of hydration. The result of the addition of a very strong solution of such a salt to a hydrophilic sol is a two-fold one. A small initial amount of the added electrolyte discharges the micelles. Addition of further increments of an electrolyte eventually brings about dehydration of the micelles due to the great attraction of the added ions for water, or to the effect of the solute in reducing the diffusion pressure of the water or to both, and the resultant separation of the dispersed phase out of the system. Salting out, therefore, consists in an electrical discharge of the micelles, followed by their dehydration. It results in the destruction of both of the stability factors of the system.

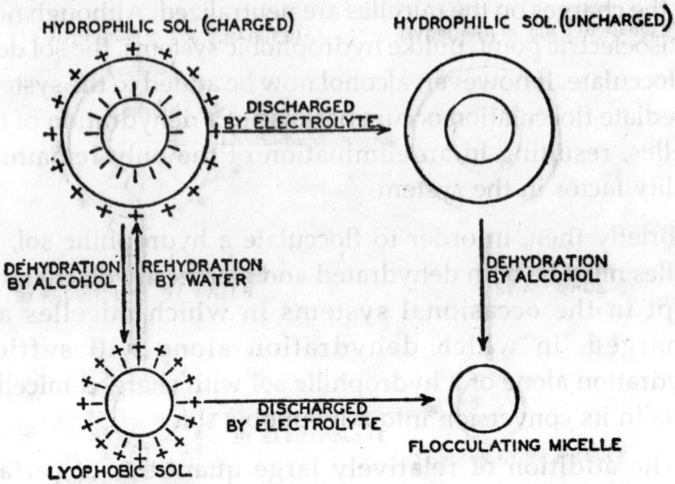

Fig. Diagrammatic representation of the flocculation of a micelle of a hydrophilic sol.

AMPHOTERIC PROPERTIES OF PROTEIN SOLS

Protein sols differ from most others in that the micelles are amphoteric, i.e. they may act either as an acid or as a base. The acid properties of proteins depend upon their COOH groups; their basic properties upon their NH_2 groups.Whether the proteins will combine with acids or bases depends principally upon the pH of the dispersion medium. In a gelatin sol, for example, in which the pH of the medium is above the isoelectric point the COOH groups of the molecules react with a base such as sodium hydroxide, forming sodium gelatinate. This compound then dissociates into sodium ions and negatively charged gelatin micelles. If the pH of the medium is below the isoelectric point the NH_2 groups of the gelatin molecules may combine with the molecules of an acid such as HCl forming gelatin hydrochloride.Dissociation of this compound produces chloride ions and positively charged gelatin micelles.

Like the micelles of other sols, those of proteins are uncharged with respect to the medium at the isoelectric point. Therefore no migration of the micelles occurs if an electric current is passed through a protein sol at its isoelectric point. At pH values higher than the isoelectric point protein micelles migrate towards the anode, while at values below the isoelectric point they migrate towards the cathode.

The principles governing the stability and flocculation of a protein solare similar to those which hold for other hydrophilic sols with the one further complication that protein micelles may be either positively or negatively charged. Protein sols are stable at their isoelectric point because, although uncharged, they possess, like all hydrophilic sols, micelles which are highly hydrated. On either side of the isoelectric point the micelles of a protein sol have the additional stability factor of an electrical charge. Addition of a sufficient quantity of a dehydrating agent such as alcohol to a gelatin sol at its isoelectric point will result, as it does with an uncharged agar sol, in immediate flocculation. If the micelles are charged, however, addition of alcohol will result not in flocculation, but in the conversion of the system into a lyophobic sol. Addition of a suitable electrolyte to this sol will result in a discharge ofthe particles and their consequent flocculation.

PROPERTIES OF GELS

Under certain conditions most hydrophilic sols change into gels. Any gelatin sol which is not too dilute, for example, will set upon standing and form a gel. Everyone is familiar with such gelatin gels, often as they appear on the table under the guise of desserts. Other familiar gels are the agar gels widely used as a medium for the culturing of bacteria, fungi, algae, etc., ordinary household jellies which are basically pectin gels, and starch gels. The latter also sometimes come before our eyesby the dessert route, under the name of corn starch puddings. Some hydrophilic sols, however, do not ordinarily form gels. This is true of sols of gumacacia and of some protein sols.

The gel-forming capacity of some substances is very remarkable. A gelatin sol containing as low a proportion as one part of gelatin to one hundred ofwater will usually gelate. Agar gels containing only 0.15 per cent of agar can be prepared. In such a gel one part of agar has the property of removing the liquidity from nearly 700 parts of water.

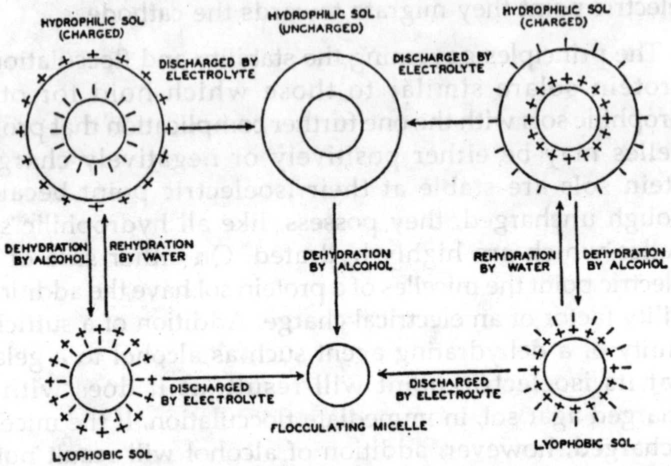

Fig. Diagrammatic representation of the flocculation of a micelle of a protein sol

Since gels form from two-phase colloidal systems, it is generally assumed that they also are two-phase systems. In all of the gels which we shall have occasion to consider, the liquid phase is water, although gels in which one component is some other liquid are well known. Gels are usually rigid enough to maintain their shape under the stress of their own weight. This means that they will be molded to the shape of the vessel in which the gelation has occurred, and will retain the shape of that vessel after being removed from it.

Solutes diffuse through gels almost as rapidly as through pure water unless the gels are very concentrated. The diffusion of a solute through a. gel is easily demonstrated by a simple experiment. A gelatin sol is allowed to solidify in a test tube which is then inverted in a shallow dish containing a solution of a dye such as methylene blue. Within 24 hours diffusion of the dye into the gelatin gel can be detected. Because of the fact that there are no convection currents in a gel to complicate the results, this is perhaps the best visual method of demonstrating the diffusion of solutes.Two equal quantities of an electrolyte, one dispersed in water, the other in a gel of equal volume, will conduct an electric current almost equally well. In other words ionic mobility is apparently as great when the ions are dispersed in a gel as when they are dispersed in pure water.The velocities of chemical reactions occurring in a gel medium are not appreciably different from the velocities of the same reactions occurring under the same conditions of solute concentration, temperature, etc., in a water medium. Neither of the last two general statements are strictly true for very concentrated gels.

The above three properties of gels distinguish them clearly from the solid or amorphous states of matter. These properties must be reconciled with any acceptable theories of the structure of gels.

ELASTIC GELS

Two general types of gels are usually recognized, the elastic type, and the non-elastic type. The best known example of the latter is the silica gel. Elastic gels are the important type biologically. Gelatin-water and agar-water gels are probably the best-known examples of this type of colloidal system. When such a gel dries a gradual and consistent shrinkage in its volume ensues until desiccation is complete. After desiccation the dry matter of the gel will imbibe water, but no other liquid. Gels in which the liquid phase was other than water will imbibe, after desiccation, only the liquid originally present.

Elastic gels are generally heat reversible. When heated such gels are converted into sols (solation) and when cooled such sols resume their gelcondition (gelation). Usually this reversal of state can occur a number of times to a colloidal system without greatly affecting its physical properties when in either the sol or gel condition. The temperatures at which the solation and gelation of a given colloidal system occur are not identical. The processes differ in this respect from the melting and freezing of a solid. For example, a 4 per cent gelatin sol gelates at about 280 C, but the resulting gel must be raised to about 31 0 C. before solation will occur. For agar gels the temperature spread between gelation and solation is much greater. The viscosity of a heat reversible sol increases steadily with a decrease in temperature and shows no sudden change as the sol passes into the gel condition.

THE STRUCTURE OF GELS

Numerous theories purporting to explain the structire of gels have been advanced, and there is no doubt that there are elements of truth in most of them. Since it seems improbable that the molecular architecture of all gels is the same, it is unlikely that any one of the proposed theories will apply equally well to all gels. The structure of gelsis too fine to be resolved by the microscope, and only rarely has the ultramicroscope revealed anything when it has been used as an instrument for studying gel structure. Necessarily, therefore, most evidence of the structure of gels is indirect.

There seems to be little doubt that in non-elastic gels of the silicic acid type that the solid phase is crystalline, and forms a sort of a rigid framework. The liquid phase is held in the interstices of this solid framework.In some respects this may be regarded as the simplest type of gel structure.One of the older theories of the structure of elastic gels is the so-called honeycomb theory. According to this theory the fluid phase of the gel is discontinuous, being separated into minute chambers or compartments,bounded on all sides by films of the solid, or at least of a more solid phase.The analogy between such a cellular structure and a honeycomb is obvious.The solid phase is not necessarily supposed to be composed of the chemically pure, originally dispersed material; it may represent simply a phase which isrelatively rich in the dispersed substance as compared with the fluid phase.In a gelatin gel, for instance, it might be supposed that the solid phase is relatively rich in gelatin but poor in water while the converse is true for the fluid phase.

At the present time an hypothesis more generally favoured is that both the solid and liquid phases of a gel are continuous. The solid phase is usually visualized as a meshwork of long, tangled fibrillae of ultramicroscopic, or sub-ultramicroscopic dimensions; the spaces within this interwoven mesh being occupied by the fluid phase. This theory is often called the brushpile theory in allusion to the supposed jumble of intermeshing threads of the solid phase.The remarks made in the preceding paragraph concerning the constitution ofthe two phases are also valid for this theory.

This last theory appears to be most acceptable in the light of the known facts regarding the very slight effect of gels upon diffusion, conductivity, and the velocity of chemical reactions. Ultramicroscopic examination of certain gels has also yielded evidence which appears to support this theory.Since elastic gels can be readily transformed into hydrophilic sols, and hydrophilic sols into gels, there is good reason for believing that hydrophilic sols may also possess a fibrillar structure. Such a structure is also postulated by many authorities for protoplasm, as later discussion will show.

HYSTERESIS

The statement is sometimes encountered that gels possess the faculty of memory. This statement is not, of course, to be accepted literally. It is merely a way of saying that the previous treatment to which a gel has been subjected have an influence, often marked, upon its behaviour, so that in an allegorical sense the gel may be said to remember those treatments. This phenomenon of the influence of the previous treatment of a gel upon its behaviour is known as hysteresis. Three gelatin gels were prepared containing respectively 10, 20, and 40 g.of gelatin per 100 cc. of water. Strips of these gels of equal rectilinear dimensions and thickness were then dried in a current of warm air until all of them were reduced to a moisture content of about 3.5 per cent.

In spite of the fact that all three of these gels possessed the same water content when immersed in water their swelling behaviour was quite different,depending in each sample on the previous history of the gel. The gel which was originally prepared in the proportion of io parts of gelatin to 100 of water swelled the most, followed in order by the one originally prepared in aproportion of 20 parts of gelatin to 100 of water, and finally by the gel originally prepared in a proportion of 40 parts of gelatin to ioo parts of water.The example cited is just one of the many hysteresis effects which have been recognized in gels.

This use of the term hysteresis should not be confused with its commonuse in another sense in physics and engineering and even the factor of time may induce such effects in gels. It follows that in experimental work with gels, if results are to be comparable, all the gels used in a given experiment or set of experiments must have had identical previous histories.

SYNERESIS

Syneresis may be defined as the spontaneous separation of aportion of the liquid component of a gel. The liquid which separates is not quite pure, however, being essentially a very dilute sol. Syneresis is a widely observed phenomenon. It may be observed in both gelatin and agar gels.The so-called bleeding of agar culture media is familiar to all who work with them. The liquid which usually forms around gelatin gels which have stood for some time is often a result of syneresis. It has been suggested that several important biological phenomena involve the process of syneresis.These include the formation of vacuoles in plant and animal cells, the separation of serum from a blood clot, the exudation of serum into a blister, glandular secretion, and the contraction of muscles.

THIXOTROPY

If a trace of sodium chloride is added to a test tube full of 1o per cent bentonite (a colloidal clay), and vigorously shaken, a colloidal sol will result which will set to a gel after standing a few minutes. By shaking, this gel can be converted to a sol which will again set upon brief standing. The process may be repeated an indefinite number of times. This phenomenon is called thixotropy. Protoplasm also exhibits thixotropic reactions. Stirring of the protoplasm or subjection of a cell to pressure can beshown to greatly reduce its viscosity and probably induces gel to sol changes.Thixotropic phenomena may therefore play important roles in cellular physiology.

Chapter 6

Plant Cells

THE STRUCTURE OF PLANT CELLS

The typical cell of the higher plantsis a tiny compartment enclosed by a tough elastic wall. The wall of cells consists of two major parts:

1. The middle lamella and
2. The primary wall.

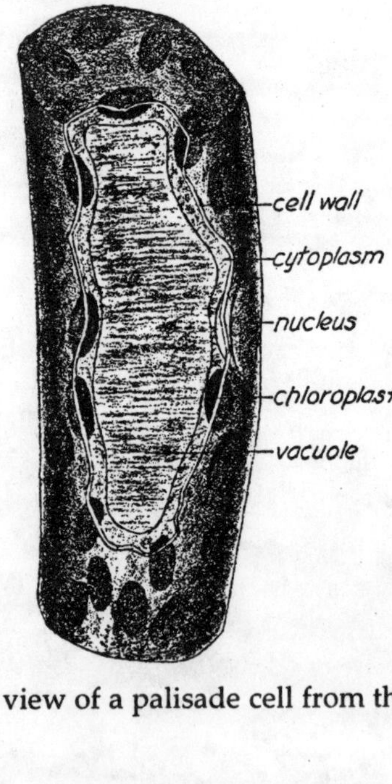

Fig. Perspective view of a palisade cell from the leaf mesophyll

In walls of many plant cells a third structural component, the secondary wall, is also present. Although some plant cells are known which do not have a well defined cell wall, this structure is so generally present in plant cells as to be considered one of their characteristic features. Lining the interior of the wall and occupying more or less of the cell cavity-nucleus is the protoplasm. The protoplasm of active cells is a transparent, slightly viscous, granular material that any conspicuous structural background. It is not homogeneous, however, and vacucle contains a number of definite structures. One of these, the nucleus, is a denser body which is more or less pheroidal in shape and is separated from the remaining protoplasm by a definite membrane, the nuclear membrane. Perspective view of a rounding the nucleus are:

1. A clear palisade cell from the leaf mesophyll. liquid known as the nuclear sap,
2. A delicate network of denser material, the reticulum, and
3. One or more small spherical masses of material known as the nucleolus or nucleoli.

All of the protoplasm outside of the nucleus of the cell constitutes the cytoplasm. In a typical mature plant cell the cytoplasm is present as a thin layer lining the inner surface of the cell wall. The two boundary layers ofthe cytoplasm that in contact with the cell wall and that in contact with vacuole are called the cytoplasmic membranes. Imbedded in the cytoplasm are numerous well differentiated bodies known as plastids. Plastids are specialized cytoplasmic structures which are usually centres of certain types of physiological activity. They are commonly classified on the basis of their colour into three groups: The leucoplasts which are colourless, the cizioroplasts which contain the green chlorophyll pigments (also yellow pigments), and the chromoplasts which contain red or yellow pigments. Chondriosomes, minute rod-like or granular bodies, are also found in the cytoplasm.

The significance of these structures is not positively known, although a number of different roles have been ascribed to them.

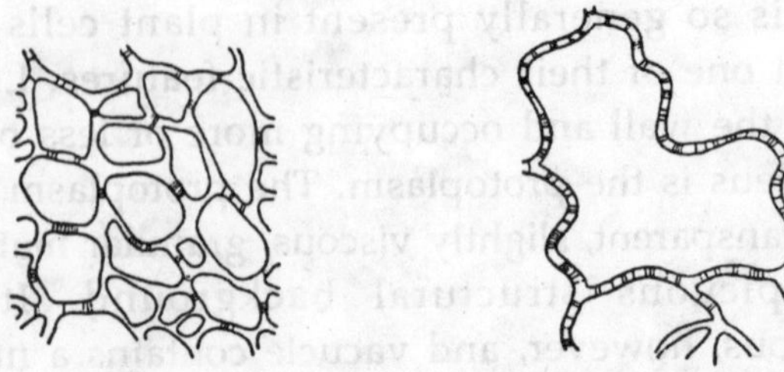

Fig. Plasmodesms in cell walls of tobacco: (A) sieve tubes and companion (B) epidermal cells of leaf.

Although the cell wall appears to imprison each protoplast, and to effectively isolate it from the protoplasm of adjoining cells, actually there is probably a continuation of protoplasm from cell to cell. By certain techniques it can be demonstrated that minute pores extend from cell to cell through the cell walls. These pores often contain cytoplasmic strands which connect the cytoplasms of adjacent cells. These strands are termed plasmodesms.

Livingston has demonstrated the occurrence of plasmodesms in the walls of cells from a number of different tissues of the tobacco plant, and it is generally supposed that they are of widespread if not universal occurrence inplant cell walls.

The bulk of the interior of mature plant cells is occupied by a single large cavity, the vacuole, which is filled with cell sap. The cell sap is composed of water in which a great variety of substances are dissolved or colloidally dispersed.

There is no general agreement among cytologists regarding the exact classification of the parts of plant cells. The vacuole, for example, is frequently classified as a part of the protoplasm because it first appears as minute droplets in the protoplasm of very young cells. Physiologically, however, the vacuole of the mature cell is as distinct an entity as the protoplasm or cell wall and it is therefore considered as a separate part of the cell in this book.

The following classification includes the principal parts of a mature plant cell. A few of these parts, as for example the various kinds of plastids, do not occur in every cell. Middle lamella, Cell Wall, Primary Wall, Secondary Wall, The Mature Plant Cell, Protoplasm, Cytoplasm, Nucleus, Vacuole, Plasmodesms, Cytoplasmic membranes, Undifferentiated cytoplasm, leuco plasts, Plastids, chloroplasts, chromoplasts, Chondriosomes nuclear membrane, Nuclear sap Reticulum Nucleolus Water, Various compounds in solution or in a state of colloidal dispersion.

THE ORIGIN AND DEVELOPMENT OF CELLS

As soon as it became clear that all plant and animal tissues were composed of cells the question of the origin of cells naturally arose. This proved to be a difficult problem for the pioneer investigators and for many years there was much disagreement over the question. The now universally accepted principle that cells can arise only by the division of pre-existing cells was first demonstrated beyond any reasonable doubt by Niigeli just before the middle of the Nineteenth Century. A number of methods are now known by which this division is accomplished but a detailed discussion of them is beyond the scope of this book. In higher plants cell division occurs chiefly in certain restricted regions called meristems. The production of new cells involves not only the division of pre-existing cells, but the subsequent enlargement and maturation of their cell progeny.

THE FORMS AND SIZES OF CELLS

All newly formed cells do not differentiate morphologically in the same way. Some elongate parallel to the axis of growth more than in other directions and thus produce the longer fiber cells that make up much of the xylem and phloem tissues. These cells commonly develop walls that are greatly thickened. Some cells enlarge about equally in all directions forming isodiarnetric cells, the walls of which are never greatly thickened. Cells of this type are present in the pith, leaf mesophyll and in other parenchymatous tissues. According to Lewis both plant and animal cells are basically tetrakaidecahedrons; i.e., 14 sided although many other geometrical shapes are found, especially in specialized types of cells. An almost endless variation in cell shapes and sizes may be seen in the tissues of any vascular plant, all of which may develop from similarly shaped meristematic cells.

In size cells show an equally great range. Most plant cells have diameters that fall somewhere between 10 and 100. A single cubic centimeter oftissue may, therefore, contain millions of cells. Some cells, however, are much larger. Such cells have lengths that are many thousand times their diameters.

THE CELL WALL

One of the most important features of plant cells is the presence of a conspicuous cell wall that encloses the living protoplasm. There are great variations in the thickness of the walls of different kinds of cells and even greater differences in the physical and chemical properties of cell walls. In fact the walls of plant cells exhibit such great contrasts in structure and chemical composition that it is difficult to single out any specific cell wall as having a structure that may be considered typical of plant cell walls in general. There are, however, certain features of the cell walls of the vascular plants that are almost invariably present. Probably the most important of these is the almost universal presence of cellulose as the structural framework of the wall. A second feature of plant cell walls is that the cellulose seems invariably to possess a crystal-like structure. No plant cell wall however, is composed solely of cellulose. There is present in every wall greater or lesser quantities of one or more other substances in addition to cellulose. The most important of these are the pectic compounds, lignin,hemicelluloses, cutin, and suberin. A third characteristic of most cell walls is their lamellate structure.

Most cell walls seem to be an aggregation of numerous delicate lamellae of varying physical and chemical properties, all of which are firmly welded together forming the superficially homogeneous wall.

ORIGIN AND DEVELOPMENT OF THE CELL WALL

In the vascular plants the division of a plant cell is preceded by the division of the nucleus. Nuclear division is a complicated process involving organization of the nuclear reticulum into unit chromosomes, splitting of those chromosomes, migration of one of the chromosomes of each pair to opposite ends of the cell, and finally, reconstitution of a daughter nucleus from each set of daughter chromosomes. Just after the final stage (telophase) of mitosis a membrane develops in the centre of the cell and extends its margins until it makes contact on all sides with the existing cell wall, thus forming a septum which separates the protoplasts of the newly formed cells. This first membrane, the middle lamella,. does not contain cellulose in quantities that permit its detection but seems largely, if not entirely, composed of colloidal pectic compounds. During the enlargement of the cell a thin layer composed largely of cellulose is deposited from the cytoplasm of each young cell on the two sides of the middle lamella. This primary wall contains large amounts of pectic materials in addition to cellulose and may vary in thickness and in physical properties at different times of the year. Primary walls may become considerably thickened as is the case in the collenchyma cells of young stems. As long as a cell wall possesses a marked elasticity and exhibits reversible changes in thickness or in other physical properties it is classed as a primary wall. The walls of cambium cells, parenchyma cells and collenchyma cells are all examples of primary cell walls.

After the enlargement of the cell ceases the deposition of cellulose on the inner surface of the primary wall may continue until the wall becomes conspicuously thickened. When, as in most cases, this thickening is accompanied by an almost complete loss of the elasticity of the wall the added material is known as the secondary wall. Secondary walls are further characterized bythe absence of reversible changes in thickness and by their low content of pectic compounds. They are largely composed of cellulose which is frequently associated with lignin, hemicelluloses, or other membrane substances. In extreme cases secondary thickening may continue until the wall occupies most of the interior of the cell. The formation of a secondary wall prevents any further enlargement of the cell. All cells have a middle lamella and a primary wall, but secondary walls are present only in certain types of cells.

One of the newly formed cells has undergone considerable enlargement. Phloem fibers, stone cells, tracheids, and wood fibers are typical examples of cells with prominent secondary walls.

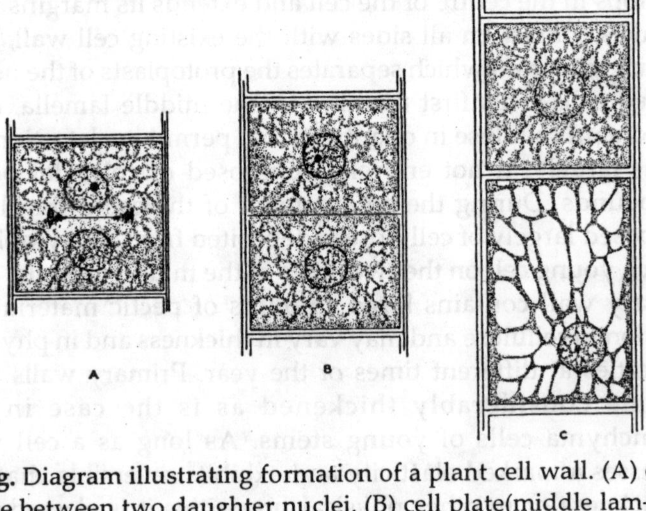

Fig. Diagram illustrating formation of a plant cell wall. (A) cell plate between two daughter nuclei, (B) cell plate(middle lam-ella) completely separating the daughter cells, (C) primary wall has been formed on each side of the middle lamella. One of the newly formed cells has undergone considerable enalrgement

Increase in thickness of the wall usually appears to take place by the addition of definite layers of cellulose or other cell wall constituents to the inner surface of the existing wall. In most walls these layers are too thin to be detected without swelling or otherwise treating the wall.

THE STRUCTURE OF THE CELL WALL

The basic unit in the structural organization of the cellulosic framework of the plant cell wall is the cellulose molecule. Cellulose molecules are not units of definite molecular weight but long chains of varying length formed by the condensation of at least a 100-120 and possibly many times this number of /3 d-glucose molecules1. Studies of cellulose walls with the X-ray and polarized light have furnished evidence that the molecules of cellulose are aggregated into bundles known as micelles. The micelles have been estimated upon the basis of X-ray photographs to have a diameter of approximately 5 or 6 mp and a length of about 6o m. A unit of this size would consist of about sixty parallel cellulose chains each being made up of about 120 glucose units. It is probable that in many cell walls long chain-like molecules formed by the condensation of other sugars are associated with the cellulose chains in the micelles.

Originally the micelles were believed to be well defined units which were cemented together by some non-crystalline material. Recent work indicates however, that many of the cellulose chains are much longer than the micellar aggregates and extend from one micelle to another, thus welding the micelles into a coherent an astomosing system. It is also probable that the micellar aggregates vary in size. Cellulose walls are no longer regarded, therefore as a system of discrete crystalline units bound together by some cementing substance but rather as a structure composed of molecular aggregates are welded together by interlocking cellulose molecules.

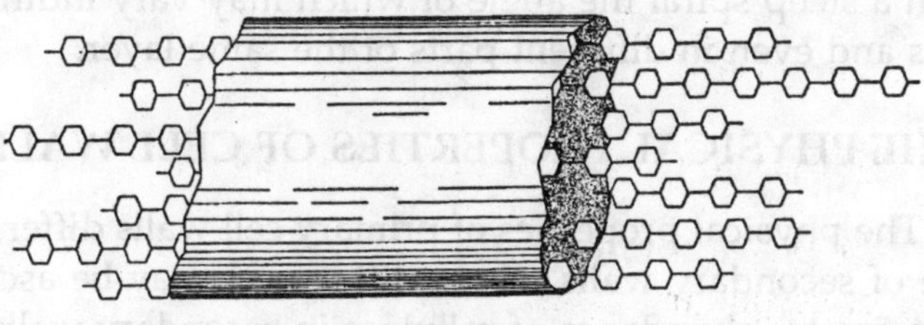

Fig. Diagram illustrating structure of a small portion of a cellulose micelle.

There seems to be little doubt regarding the presence of intermicellar spaces. The spaces form an interconnecting system between the anastomosing micelles.

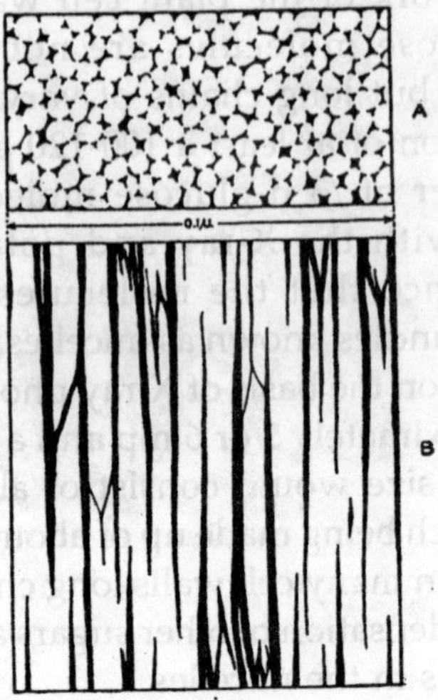

Fig. Micellar structure of walls. Black areas represent intermicellar spaces; white areas represent the cellulose micelles. (A) cross section, (B) longitudinal section.

In walls that are almost pure cellulose, such as the secondary wall of cotton fibers, it is possible that the intermicellar spaces are filled chiefly with water.The smallest visible units of cellulose walls are delicate thread like strands or fibrils. In primary walls these fibrils form a loose anastomosing network the meshes of which are usually filled with colloidal pectic compounds. In secondary walls the fibrils are often grouped into coarser strands which wind around the cell in a steep spiral the angle of which may vary indifferent layers and even in different parts of the same layer.

THE PHYSICAL PROPERTIES OF CELL WALLS

The physical properties of primary cell walls differ from those of secondary walls. These differences may be ascribed to the greater abundance of cellulose in secondary walls and to differences in the structural organization of the cellulose in the two kinds of walls. Both primary and secondary walls are transparent to wavelengths of the visible spectrum and both are usually quite pcrmeable to most substances dissolved in water that compare favourably with those of steel. The breaking strength of a flax fiber, for example, may be as great as 110 kg. per mm.2 of wall area while the tensile strength of a hardened spring steel ranges between 150 and I 70 kg. per mm2. Although the tensile strength of primary walls is considerably less than that of secondary walls it is only rarely that they are subjected to strains in excess of their breaking strength.

Primary cell walls are usually quite elastic. The mesophyll cells of the leaves of some species, for example, are known to undergo reversible changes in volume of 30 per cent or more in response A to changes in turgor pressure. Secondary walls, on the other hand, are elastic only within very narrow limits and break suddenly when their tensile strength is exceeded. The elasticity of primary walls is undoubtedly related to the loose open mesh work of cellulose strands that compose it, while the close parallel orientation of the cellulose micelles in secondary walls prevents any appreciable elongation without rupture.

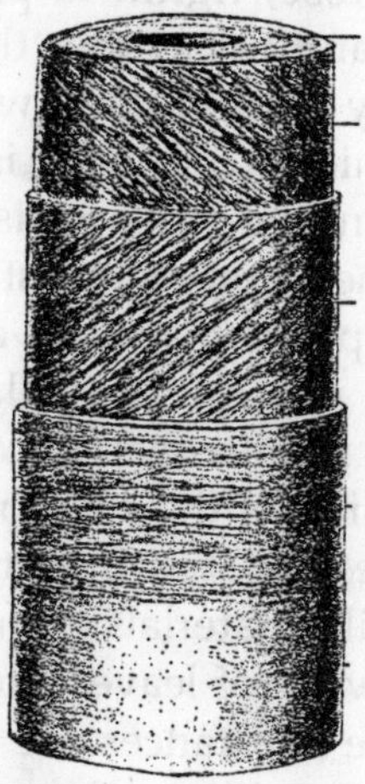

Fig. Diagram illustrating the structure of a thickened plant cell wall.

CHEMICAL CONSTITUENTS OF CELL WALLS

Cellulose is by far the most abundant compound found in the cell walls of the higher plants. Associated with cellulose in all the primary cell walls of vascular plants are greater or lesser amounts of pectic compounds. Their presence in the middle lamella and in the intermicellar spaces of the primary walls has already been noted. Lignin is an important constituent of the walls of most of the cells that make up woody tissues and it also occurs commonly in other thickened walls. Although lignin can be isolated from cell walls by suitable treatments as a brownish amorphous substance it is probably altered considerably in the process. Its structural formula and molecular weight are unknown. Since the lignins produced in different species appear to differ in their chemical properties it seems very probable that lignin is not a compound of definite molecular weight but a complex mixture of a number of chemically similar substances. The lignin of spruce wood has been extensively studied and several workers have suggested that its molecular weight lies between 800 and 900.

Lignin first appears in the middle lamella and primary wall and later canbe detected in the secondary wall. When associated with cellulose, lignin is present in the spaces between the micelles and not within the micelles. Lignifled walls are usually freely permeable to water and solutes. The tensile strength of lignified cell walls is the same as that of cellulose walls but lignified walls resist compression better than cellulose walls. The increased resistance of lignified walls to compression is explained by the assumption that the presence of lignin in the intermicellar spaces welds the cellulose micelles into a single coherent mass and thus prevents the bending and buckling of the cellulose strands when they are subjected to compression strains. Cutin is the name applied to the mixture of wax-like materials found on the outer surface of the epidermal cell walls of leaves, stems, fruits, and other organs. These wax-like substances are intimately associated with cellulose and often with pectic substances producing a wall of great structural complexity. Cutinized cell walls are relatively impermeable to water. The presence of cutin in the outer walls of epidermal cells greatly reduces the evapouration of water from the surfaces of plant tissues. Subcrin is similar in many of its properties to cutin. It constitutes an important part of cork cell walls and it is also found in the walls of a few other specialized types of cells. Most of the surface of perennial plants, aside from the leaves and very young stems is covered with suberized cell walls. Such walls are relatively impermeable to water.

Hemicelluloses are a poorly defined group of poly saccharides associated with cellulose in plant cell walls. They are not chemically related to cellulose, as the name implies, but possess very different chemical and physical properties.

Callose is the name given to a carbohydrate membrane substance found in the perforated septa (sieve plates) of the sieve tubes. Similar material presumably of the same chemical composition, has been found in pollen grains and constitutes the inner layer of pollen tubes. It has also been reported as occurring in the fungi. The exact chemical composition of callose is unknown since it has never been obtained in sufficient amounts to permit quantitative determinations.

Chitin, a nitrogenous substance common in the exoskeleton of insects, is a constituent of the walls of many fungi and bacteria. It has been reported as being present in the walls of certain algae but it is unknown in any of the higher plants.

Tannins are commonly found in the cell sap but they also occur in the walls of certain tissues, especially cork and wood cells.

Mucilages are common constituents of the outer walls of many water plants and occur also in the outer walls of some seed coats, in glandular hairs and in other specialized tissues.

Inorganic compounds such as silica and salts of calcium, iron, and other metals are also present in some plant cell walls. None of these inorganic compounds, however, are regarded as essential constituents of the cell wall.

PROTOPLASM

In most mature cells of the vascular plants protoplasmis present only as a thin layer covering the inner surface of the cell walls but in some specialized cells branching strands of protoplasm also extend across the vacuole. Under high magnification the protoplasm of active cells appears as a colourless fluid, in which are suspended numerous tiny granules and droplets of insoluble materials. These granules frequently exhibit active Brownian movement. The fluid component of protoplasm also is frequently in motion, streaming around the inner surfaces of the cell walls. Embedded in this fluid component, and carried passively by it, are the specialized bodies known as plastids.

Although protoplasm appears to be a simple liquid, no simple liquid could possibly possess the remarkable powers of synthesis, assimilation, reproduction growth and sensitivity that characterize the protoplasm of living plant cells.The properties and behaviour of protoplasm clearly show that it is not a substance but that it must be regarded as a complex system of substances. This system is dynamic; it is constantly undergoing changes yet at the same time the changes are so regulated and controlled that the system is not disrupted. A cell is alive only so long as the organization of this dynamic protoplasmic system is maintained.

The protoplasmic system of living plant cells almost invariably contains a well differentiated globular body known as the nucleus. All of the protoplasm outside of the nucleus is designated as cytoplasm. Although the nucleus is separated structurally from the cytoplasm by a delicate membrane the two components are not physiologically isolated. The nucleus and the cytoplasm appear to be mutually dependent and there is good reason to believe that the nucleus controls and regulates the physiological processes that occur in the cytoplasm. Probably because of the close relationship between the nucleus and the cytoplasm many biologists use the terms protoplasm and cytoplasm synonymously. In this book, however, the terms will be used as has been indicated earlier in this chapter.

THE CHEMICAL COMPOSITION OF PROTOPLASM

Since protoplasm is a dynamic system of substances it is not possible to subject it to chemical analysis without destroying it. In the strict sense, therefore, it is impossible to discover the chemical composition of protoplasm. It is possible, however, to examine the substances present after the protoplasmic system has been destroyed and to determine their chemical composition and relative abundance. A number of such studies have been made.

Water is the chief component of all physiologically active plant protoplasm usually making up more than 90 per cent of the system. The water content of the protoplasm of dry seeds, on the other hand, may be less than 10 per cent. Most attempts to determine the chemical composition of plant protoplasm have been made upon species of the myxomycetes. At certain stages in their life history these organisms consist of naked masses of labile protoplasm. They are often found flowing over rotten logs in damp woods. The fact that the myxomycetes provide relatively large quantities of protoplasm entirely free from cell wall material has made them a favourite object for chemical analysis. Even in such organisms, however, not all of the constituents of the plant body can be regarded as integral parts of the protoplasm. Distributed throughout the protoplasmic mass are particles of foods and other inert materials which cannot be separated from the protoplasm.

As shown in this analysis proteins and other nitrogen-containing compounds constitute the bulk of the organic matter in the plasmodium of this species. Many different varieties of proteins are known to occur in the protoplasm of plant cells. They are compounds of enormous molecular weight and undoubtedly make up a large proportion of the labile structural framework of the protoplasm.

Lipids constitute a smaller fraction of the protoplasm than the proteins. Three types of lipids occur in the protoplasm; the true fats (oils), the phosphatides (phospholipids) and the sterols, of which phytosterol is an example. The oils are generally suspended in the protoplasm in the form of minute globules. They are probably more important as food reserves than as actual constituents of the protoplasm. The phospholipids and sterols, on the other hand, are believed to be essential constituents of the protoplasmic system.

The water-soluble carbohydrates, amino acids, etc., present in the plasmodium of this species are probably almost entirely foods. The inorganic compounds (mineral matter) in plant cells are chiefly the phosphates, chlorides sulfates, and carbonates of magnesium, potassium, sodium, and calcium.

It supplies no more information regarding the organization of protoplasm, however, than a chemical analysis of the ground up debris of a wrecked house would furnish regarding the structure of that house. This point is worthy of emphasis, since with the progress of biology it becomes clearer and clearer that the properties of protoplasm are as much a function of its physiochemical organization as of the specific kinds of compounds present. Although we commonly refer to protoplasm as the essential constituent of all living cells, it is evident that there are at least as many different varieties of protoplasm as there are species of plants and animals. All of them, however, are dynamic systems composed of the same types of compounds, and all of them possess a colloidal organization of a complex type.

THE PHYSICAL PROPERTIES OF CYTOPLASM

Although it is not possible to explain adequately the mechanisms responsible for the physiological activities of cytoplasm it is possible to study the behaviour of cytoplasm under various Percentage of dry weight.

1. Water soluble substances, chiefly from vacuoles: Monosaccharides 14, 2 Proteins 2. 2 Amino acids, asparagine, etc 24.3
2. Insoluble organic substances, principally constituents of the protoplasm: Nucleo-proteins 32.3 Nucleic acids 2.5, Globulin 0.5, Lipo-proteins 8, Neutral fats 6.8, Phyto stero l3.2, Phosphatides I.3, Other organic matter 3.5.

3. Mineral matter, about half water soluble 4.4 conditions. Studies of this kind have furnished considerable information about some of its physical properties.

TRANSPARENCY

Like the walls of most living cells protoplasm is usually transparent to the wavelengths of the visible spectrum. Photochemical reactions can therefore occur in cells located well below the surfaces of many plant organs such as leaves.

ELASTICITY

Some authorities consider cytoplasm to be highly elastic. Seifriz has shown that it is possible to draw out the cytoplasm of living plant cells into long threads which behave as elastic bands and snap back into the cytoplasmic mass when released. It is difficult to determine, however, whether the elastic properties of such cytoplasmic threads are characteristic of the cytoplasm as a whole or only of the membranes which enclose it. The phenomenon may be due to surface tension forces rather than to a true elasticity of the cytoplasm proper. Experiments designed to measure the elasticity of the cytoplasm within its limiting membranes indicate very low values. Probably active cytoplasm as a whole possesses elastic properties because of the presence of elastic surface membranes but the bulk of the cytoplasm which lies within these membranes behaves as a liquid and is essentially non-elastic. The entire protoplasm of a cell may become elastic, however, when the viscosity is greatly increased, particularly when it is in the gel state.

VISCOSITY

Numerous ingenious methods of measuring the viscosity of cytoplasm have been devised but the results obtained are not in agreement. The one conclusion which emerges most clearly from all of these studies is that the viscosity of the cytoplasm of living cells may vary greatly. It has also been found that not all of the cytoplasm in one given cell has the same viscosity at the same time. The surface layers of the cytoplasm are more viscous than the interior cytoplasm. In general cells that are physiologically active have cytoplasm of low viscosity while in dormant cells the cytoplasm may be as stiff as a rigid gel. According to Seifriz fluid cytoplasm usually has a viscosity ranging from 800 (the consistency of glycerine) to 8000 (the viscosity of a thick sugar syrup) times that of pure water. On the other hand Heilbrunn considers active cytoplasm to have a viscosity only four or five times that of water. Probably most workers would consider the latter estimate a closer approximation to the facts than the former. The viscosity of the cytoplasm in active cells may change rapidly in response to mechanical injury or electric shock, to changes in temperature differences in acidity, and exposure to various chemical compounds. Dehydration increases the viscosity of the cytoplasm and death of the cell results in a marked increase in viscosity.

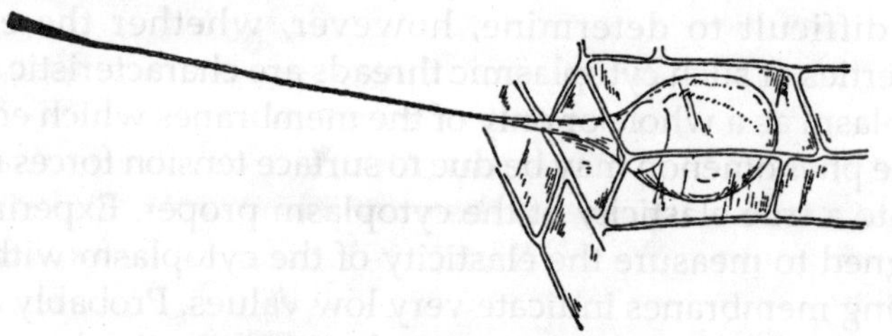

Fig. Demonstration of the "elastic" property of cytoplasm. Redrawn from Seifriz (1936)

IMMISCIBILITY WITH WATER

The cytoplasm of cells that are physiologically active is invariably composed predominantly of water, yet when the protoplasts are extruded from such cells into an aqueous medium they do not ordinarily mix with the water. The failure of cytoplasm to become dispersed through the water is largely, if not entirely, due to the presence of a surface membrane containing fat-like constituents which are insoluble in water. When this surface film is punctured a new membrane quickly forms across the broken surface. If substances are present that prevent the development of this surface film the cytoplasm of active cells readily disperses in the water.

GELATION

Most of the cytoplasm of metabolically active cells exhibits the properties of a hydrophilic sol. In dormant or inactive cells, however, the cytoplasm apparently is often in a gel condition. No sharp line of division can be drawn between hydrophilic sols and gels and every gradation may existing a given system between a well marked sol state and a well marked gel state. Every such gradation in physical condition may also exist in cytoplasm. It may range in physical state from a highly fluid sol through a viscous sol condition to a stiff elastic gel.

COAGULABILITY

The cytoplasmic system of most physiologically active cells is destroyed by temperatures of 600 C. or above. Death of plant cells by such temperatures is generally considered to result from a coagulation of some of the protein aceous constituents of the protoplasm. A number of other factors may bring about coagulation of the protoplasm, at least in the cells of certain species. Among these are certain electrolytes electric currents, freezing, mechanical pressure, and certain wave lengths of radiant energy (especially ultraviolet radiations, X-rays, and radium radiations).

ELECTRICAL PROPERTIES

Numerous attempts have been made to determine the isoelectric point of cytoplasm. It was soon discovered that cytoplasm does not have a definite isoelectric point but rather an isoelectric range This is doubtless due to the fact that cytoplasm is a complex sol composed of many different proteins, each one of which may possess an isoelectric point different from that of the others. Measurements of the cytoplasm of cells in the root tips of different plants showed the isoelectric range to be from pH 4.6 to pH 5.0. These values appear to be representative for cytoplasm in general.

Cytoplasm is usually on the alkaline side of its isoelectric range and therefore we would expect to find that its constituent micelles are negatively charged. Cataphoretic migration of the granules in the cytoplasm of root hair cells and epidermal cells toward the anode was shown to occur, thus indicating that these particles carry a negative charge.

The proteins present in the nucleus apparently possess different isoelectric points from those of the cytoplasm. Furthermore, it is probable that some portions of the cytoplasm possess isoelectric points different from those of other parts of the cytoplasm in the same cell.

Cytoplasm contains dissolved electrolytes and would therefore be expected to conduct an electrical current. Brooks determined the electrical conductivity of the plasmodium of the myxomycete Biefeldia maxima found it to be equivalent to that of a 0.00145 N solution of NaCl, which is a relatively low value. The solution in the moss substrate on which the plasmodium was growing had a conductivity only about one-third as great as that of the plasmodium. Evidence was also found that the conductivity of the protoplasm varies according to the conductivity of the medium with which the cells are in contact.

STREAMING

In many cells the cytoplasm may be seen in active movement. In the simplest cases this movement consists of a rotation of the cytoplasm around the inner surface of the cell wall. Where cytoplasmic strands extend through the vacuole as in the case of cells of Tradescantia stamen hairs the circulation of the cytoplasm may become very complex. The fluid like portion that is in motion is often bounded by thin layers of non-moving cytoplasm. The cytoplasm immediately adjacent to the cell wall and that which bounds the vacuole often does not move. The plastids and the visible granules are carried passively around the cell by the moving cytoplasm. The causes of cytoplasmic streaming (cyclosis) are unknown. It is accelerated by increases in temperature up to the point where injury appears and checked by low temperatures, ceasing at temperatures slightly above the freezing point. Cyclosis is also stopped in the absence of oxygen and by anaesthetics in relatively high concentrations. In dilute concentrations toxic substances such as copper sulfate and narcotics accelerate the streaming movements. Light also appears to increase the rate of streaming under certain conditions.

THE PHYSICAL STRUCTURE OF CYTOPLASM

Any satisfactory explanation of the structure of cytoplasm must account not only for its physical properties and for its dynamic behaviour, but must also explain how the innumerable diverse physical and chemical reactions characteristic of living cells can occur side by side in the same general medium. The highest magnifications reveal no evidence of any structural background in active cytoplasm, yet its complex activities suggest that it must possess an intricate structural organization. The marked imbibitional capacity of cytoplasm, its stability toward electrolytes, its electrical properties, its viscosity, its coagulability and its gel-forming capacity all suggest that it is to be classed with the hydrophilic colloids. All modern students of cytoplasm are agreed that it is a hydrophilic colloidal system of extreme complexity and variability.

As was pointed out in the preceding chapter elastic gels are generally considered to possess a submicroscopic structure of long interlacing fibrillar units which hold a liquid phase in their irregular interstices. The same structure has been suggested for some hydrophilic sols. There is some evidence that the structure of cytoplasm is similar. The cytoplasm in many living cells appears fibrous and long delicate strands may extend through the vacuole of the cell. Thread-like strands likewise appear in cytoplasm when it is being manipulated with very fine glass needles. The sudden changes in viscosity that living cytoplasm may undergo when disturbed, strongly suggest a fundamental structure similar to that of elastic gels. Furthermore, living protoplasm can itself flow slowly through small openings such as those in a fine sieve without injury, but cannot be forced through somewhat larger holes without serious injury or death. This behaviour has been interpreted as evidence that a fiber-like structure exists in protoplasm; the fibers being able to slip through the holes when it is slowly streaming but being crushed and destroyed if a mechanical pressure is utilized to force the protoplasmic mass through the openings.

On the basis of this and other evidence living cytoplasm is presumed by some workers to have a skeletal structure of long submicroscopic fibrils. A fluid phase in which are suspended the visible droplets and granules is believed to be held in the interstices between the fibrils. Solutes may be presumed to be present in both phases. The whole system of fibrils and liquidis assumed to be dynamic and undergoing constant change, in which gel-like fibrils are rapidly transformed into sols and the sols form fibrils with equal rapidity. Such a structure will account satisfactorily for many of the properties of cytoplasm. For example, it offers an explanation for the rapid changes in viscosity that are known to occur in cytoplasm, it accounts satisfactorily for its hydrophilic properties, and it suggests a possible solution to the perplexing problem of how so many diverse processes can go on in the same cytoplasm at the same time. Many of these reactions may occur at the interfaces between the various fibrillar units and the liquid constituents of the system. Since different fibrillar units are undoubtedly of different chemical composition, different chemical and physical reactions could take place some at one interface and some at another. Furthermore, it is possible that some reactions occur within the liquid component and others within the more solid fibrillar units of the cytoplasm.

THE PLASTIDS

All living cells of the higher plants contain prominent cytoplasmic bodies known as plastids. These structures are usually ellipsoidal in shape and are frequently conspicuous because of the presence of pigments. The colour of the pigments in the plastids has been used as the basis for their classification but this system is highly artificial since a single plastid may be colourless, green, red or yellow at different periods of its existence.In meristematic and embryonic cells the plastids first appear as very tiny granules in the cytoplasm that range in size down to the limit of visibility. These granules gradually enlarge and differentiate untilmature plastids are produced. In growing and in mature cells plastids frequently multiply by simple fission.

In the algae the chioropiasts exhibit a wide range of size and shape but in the higher plants they show a remarkable similarity. The mature chioroplast of the higher plants is typically a slightly flattened ovate spheroid with the longer axis ranging between 4 and 6 in length. The number of plastids is not constant in different cells nor in the same cell at different stages of its development.

The structure of chloroplasts has been more thoroughly investigated than that of other kinds of plastids. Most investigations indicate that the chloroplast consists basically of a proteinaceous matrix or stroma which is probably surrounded by a membrane. In at least some chloroplasts the chlorophyll apparently occurs in small disk-shaped granules called grana which are distributed throughout the stroma. The chlorophyll is apparently associated with both proteins and lipids. The yellow pigments carotene and the xanthophylls also occur in the chloroplasts. The chromoplasts contain red or yellow pigments and are frequently very different in size and shape from the chioroplasts. Usually they are very slender spindle-shaped or needle-shaped bodies. They occur both singly and grouped in bundles. The irregular angular outline of these plastids contrasts sharply with the regular curved surface of the chloroplasts. In some species however, chloroplasts may develop red or yellow pigments and completely lose their green colour. The red and yellow colours of some fruits, notably members of the Solanaceae, and some flowers are caused by chromoplasts.The colourless leucoplasts are usually present in the cells of meristematic tissues in which they often represent juvenile stages in the development of chloroplasts and chromoplasts. In tissues not exposed to light they remain as colourless plastids and are the structures in which starch grains are formed. Specialized plastids known as elaioplasts have been described as occurring in the cells of some species. These plastids appear to be centres of oil formation.

THE NUCLEUS

The nucleus is a conspicuous spheroidal body which is imbedded in the cytoplasm. In most plant cells the nucleus has a diameter which falls within a range of 5 to 25. In the vascular plants there is usually only one nucleus to a cell, although in certain types of cells several may be present. The sieve tube elements are the only well known example of living cells in the higher plants in which no organized nucleus is present.

The nucleus is surrounded by a definite membrane and possesses a complicated internal structural organization. The larger portion of the volume of the nucleus is composed of a transparent, optically homogeneous sol or gel the nuclear sap, which surrounds a system of delicate, anastomosing threads the reticulum. The reticulum is not structurally homogeneous but contains irregular granules of material known as chromatin. Most nuclei also contain one or more nucleoli. Typically a nucleolus is a small spherical droplet of deeply staining protein and lipoidal material that is attached to the reticulum.The hereditary factors which influence the development of the organism are known to reside in the nucleus of the cells. It is clear therefore that the nucleus must exert a controlling influence over the physiological activities of the cell. There is some evidence that the nucleus is concerned with the production of the enzymes which catalyze many, if not most, physiological processes. Very little is known, however, regarding the exact role of the nucleus in protoplasmic activities.

THE VACUOLE

One of the most characteristic features of a mature plantcell is the presence of a large central vacuole filled with cell sap and entirely surrounded by the cytoplasm.

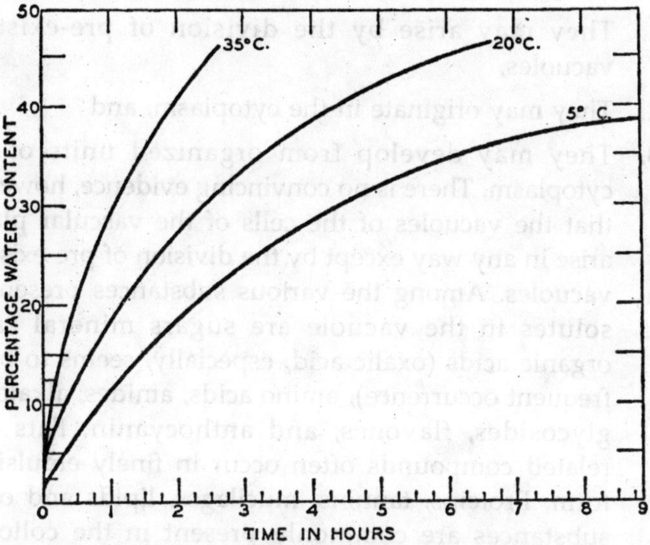

Fig. Relation beween temperature and imbibition of water by Xanthium seeds

Merismatic cells in the tips of stems and roots usually contain numerous small vacuoles scattered throughout the cytoplasm. The shape of these vacuoles varies greatly and seems to be determined by the activity of the cytoplasm. In quiescent cytoplasm the small vacuoles are usually spherical but when the cytoplasm is actively streaming they may assume many different forms. Cambium cells that are actively dividing may contain very large vacuoles. In cambium initials the vacuoles, like those in the meristems of stem and root tips, are not uniform in size or shape, but may berod-shaped, thread-like or globular. They may coalesce into a large single vacuole or they may divide into numerous smaller vacuoles. Mature cells however, whether they arise from primary meristems or from the cambium cells typically contain one large central vacuole which arises by the increase in size and coalescence of the numerous smaller vacuoles usually present in the meristematic cells.

There is no general agreement as to the method by which vacuoles originate. Three possibilities are recognized:

1. They may arise by the division of pre-existing vacuoles,
2. They may originate in the cytoplasm, and
3. They may develop from organized units of the cytoplasm. There is no convincing evidence, however, that the vacuoles of the cells of the vascular plants arise in any way except by the division of pre-existing vacuoles. Among the various substances present as solutes in the vacuole are sugars mineral salts, organic acids (oxalic acid, especially, seems to be of frequent occurrence), amino acids, amides, alkaloids, glycosides, flavones, and anthocyanin. Fats and related compounds often occur in finely emulsified form. Proteins, tannins, mucilages, lipids and other substances are commonly present in the colloidal state. Aleurone grains develop from specialized vacuoles of cells in storage tissues. Crystals of calcium oxalate are also of frequent occurrence in the vacuoles of mature cells.

HYDROGEN ION

CONCENTRATION OF PLANT CELLS

The hydration and viscosity of the protoplasm, the permeability of the cytoplasmic membranes, the activity of enzymes, the chemical activity of various ions in the cell, and various other physiological processes and conditions are all influenced more or less by the hydrogen ion concentration of the protoplasm and the cell sap. As soon as its significance was appreciated attempts were made to determine the hydrogen ion concentration of plant cells. Most of the earlier determinations were made on the juice pressed from plant tissues. Such a crude method provides only the roughest sort of an indication of the hydrogen ion concentrations in individual plant cells. The death of the cells and the mixing of the cell contents during the extraction process undoubtedly result in marked changes in hydrogen ion concentration. Such determinations are probably more nearly a measure of the pH of the cell sap than of the protoplasm. The values for expressed plant saps mostly fall within a range of pH 3.Q to pH 7.0 Direct measurements of the hydrogen ion concentration of the protoplasm and cell sap have been made by introducing indicator dyes directly into cells.By careful manipulation of a micropipette the dyes can be injected into the cytoplasm without penetrating into the vacuole or injected into the vacuole without penetrating into the cytoplasm. Only non-toxic dyes should be injected into living cytoplasm for pH determinations, else the results may be invalidated by injury or death of the cytoplasm. Dyes are considered to be nontoxic or essentially so if protoplasmic streaming continues in the same way as before the injection. In this manner it is possible to determine the reaction of the cell sap and that of the cytoplasm independently. Results of the micro injection method when applied to the root hairs of the water plant Limnobiumspongia indicated a pH value for the cytoplasm of 6.9 1 0.2. The cell sap of the same cells was found to be more acid having a pH of 5.2 1 0.2.

As indicated by the results cited above the pH of the cytoplasm and the cell sap of a plant cell may be very different. The pH of the cytoplasm of plant cells appears to be fairly constant, usually falling between 6.8 and 7.0. The pH of the cell sap is usually lower than that of the cytoplasm, values between pH 5.2 and 6.2 seeming typical for most plant cells.The cell sap of some cells, however, shows a considerably more acid reaction than this, values as low as p11 0.9 being reported for species of Begonia. On the other hand alkaline values have also been found in the vacuoles of some species. The cells appears to vary more widely in hydrogen ion concentration than the cytoplasm.

BUFFER ACTION IN PLANT CELLS

Both the cytoplasm and cell sap of plant cells are usually buffered. The former, however, as its lesser variability in pH indicates, is usually more strongly buffered than the latter. All the actual information available regarding buffer action in plants is based, however upon determinations made upon expressed plant saps. The most important buffer systems in living organisms are those composed of a weak acid and one (or more) of its salts. The principal acids which are components of plant buffer systems are carbonic phosphoric, citric, malic, tartaric, and oxalic. The first two of these acids are chemical progeny of compounds which enter the plant from its environment carbon dioxide and the phosphates, respectively. The others are products of the metabolic activity of the cells. The principal base-forming elements present in the salts which may serve as components of buffer systems are sodium, potassium, calcium, and magnesium. The buffer action of some types of plant cells is apparently due predominantly, if not almost entirely to one system. The buffering of lemon juice for example, can be accounted for almost entirely in terms of the citrate buffer system. More commonly, however, a number of different buffer systems are present in a single cell.

Chapter 7

Diffusion

The chemical elements which constitute the bulk of the body of any plant are among the commonest ones on the surface of the earth. They occur in the environment of plants as relatively simple inorganic compounds, and enter plants in the form of such compounds. From the relatively small number of compounds which enter it from its environment the green plant fabricates the numerous complex organic compounds which are essential to its continued existence as a living system.

The movement of substances into a plant from its surroundings is accomplished principally through the agency of one form or another of the process known as diffusion. Substances enter a plant partly through its aerial organs, and partly by way of its root system. From the atmosphere carbon dioxide and oxygen gases diffuse into plants, principally through the stomates.From the soil, water and the cations and anions of simple inorganic salts pass into plants by processes which are basically diffusion phenomena, although the entrance of both water and mineral salts is complicated by other factors.Similarly the loss of substances from a plant into its environment is accomplished principally by diffusion. Large quantities of water-vapour pass out of leaves and other aerial organs of plants into the atmosphere. At times oxygen gas and at times carbon dioxide gas diffuse into the atmosphere. Certain volatile compounds also escape from the aerial organs of many plants by diffusion. Similarly the roots lose carbon dioxide and other compounds into the soil by diffusion.

Likewise some of the movement of substances from one part of a plant to another is accomplished by diffusion. This is true both of the gases which move through the intercellular spaces and the water and solutes which move within the cells. However, much of the translocation of materials from one plant organ to another is accomplished by more complicated mechanisms than diffusion. Within any living plant cell diffusion of substances from one part of a cell to another is also continually in progress.

In brief, there are few if any of the physiological processes occurring in plants which do not directly or indirectly involve diffusion phenomena.

DIFFUSION OF GASES

If a small vial of bromine gas is broken under a bell jar which has previously been evacuated of air the entire jar quickly becomes filled with the brownish vapour of bromine. The distribution of the bromine gas throughout the bell jar has been accomplished by the kinetic activity of the bromine molecules, and is a simple example of the process of diffusion. If the vial of bromine be broken under a bell jar which has not been evacuated of air, the time required for the bromine gas to completely occupy the bell jar will be longer than when diffusion of the gas occurs into a vacuum. Under such conditions the freedom of movement of the bromine molecules is greatly impeded by the presence of molecules of the gases of the air, and the diffusion process is retarded. If the pressure of the air within the bell jar be increased to two atmospheres (which is equivalent to doubling the concentration of all the gases in the jar), the rate of diffusion of the bromine gas through the air would be still less than when the jar was occupied by air at atmospheric pressure.

Many other simple examples of the diffusion of gases might be cited. If a bottle of ammonia, ether, peppermint oil, or of any other readily volatile substance with a characteristic odour be opened indoors, within a very short time the distinctive odor of that substance can be detected in all parts ofthe room. Such a dispersal of gas molecules is accomplished at least in part by diffusion, although air currents often assist in speeding up such a distribution of molecules. Except in the rare case of diffusion into a vacuum diffusing molecules move between the molecules of other substances. The following some what fanciful analogy may aid in a visualization of the kinetics of the diffusion process in gases. Suppose two adjoining rooms to be connected by a closed double door. Imagine also that one of these rooms contains a large number of tennis balls travelling in various directions along straight pathways at different rates of speed. The average distances between the tennis balls are supposed to be relatively great in proportion to their diameters. Each tennis ball represents a molecule. The individual balls will be constantly bumping into each other and into the walls of the room. Because of the large number of balls present innumerable collisions will occur every second. Each time a ball strikes a wall of the room it will bounce off along a different linear pathway. Similarly, whenever two balls collide, each will be deflected out of its course along a different route, to which it will hold undeviatingly until it is again deflected from its path by another collision.The course of each ball will thus be a zigzag progression through space, each short segment of its path being terminated by a collision which changes its direction. All of the balls do not move at the same speed at any given time but the speed of each fluctuates from moment to moment as a result of the numerous collisions in which it participates. The average speed of the entire group will remain constant, however, as long as the temperature remains unchanged. As a result of this haphazard mutual buffeting, the tennis balls will remain essentially equally distributed throughout the room.

Suppose now that the double doors connecting the two rooms are thrown open. As a result of their haphazard movement some of the balls close to the door will pass into the empty room. The first ones to do this will travel without interruption until they bump into one of the walls, as their rate of progression will not be impeded by collisions with other molecules. When they hit against one of the walls of the room they will bounce back into its interior along a new pathway. As more and more of the balls pass into the originally empty room their rate of progress becomes slower since the greater their concentration in the room, the greater the number of collisions per unit of time. As soon as any appreciable number of tennis balls has invaded the empty room, as a result of their random movements, some will pass back through the doorway into the room which originally contained all of them. As long, however, as the concentration (number per unit volume) of the tennis balls is greater in one room than in the other their random movements will result in more passing through the door into the room in which their lesser concentration prevails, than in the opposite direction.In a relatively short time the concentration of tennis balls will have become equal in both rooms. In other words they have diffused from one room into the other. After equality of concentration has been established the number of balls passing through the door in one direction in any interval of time will be exactly equal to the number moving in the opposite direction. When this condition of dynamic equilibrium is attained, diffusion, in the sense the word will be used in this discussion, is no longer occurring. If, in the hypothetical illustration just described, one room were filled with white tennis balls, and the other with red tennis balls, diffusion would occur simultaneously in both directions. The red balls would diffuse toward the room in which their initial concentration was zero while the white balls would diffuse in the opposite direction. At equilibrium the concentration of the white balls would be equal throughout the two rooms, and this would likewise be true of the red balls.

Even after a dynamic equilibrium has been attained in any system haphazard kinetic activity of the molecules continues. This is sometimes referred to as diffusion, but will not be so considered in this book. The term diffusion will be used only to characterize situations in which there is gain in the number of molecules of a certain kind in one part of a system at the expense of other parts. According to this concept diffusion can only occur when the concentration of the diffusing substance is not uniform throughout the system, and the process can continue only as long as differences of concentration are maintained.

Diffusion dependent upon the kinetic energy of molecular motion and thus resulting solely from differences in concentration is often called simple diffusion in order to distinguish it from more complex types of diffusion phenomena in which other forces come into play. Examples of such more complex phenomena will be encountered in some of the following chapters. The phenomenon of diffusion is exhibited by the molecules of liquids solutes, and even solids, as well as by those of gases. Diffusion of ions also occurs, and even colloidal particles diffuse, but only at very slow rates.

Diffusion phenomena should be clearly distinguished from mass movement in which the moving units are not single molecules, but more or less extensive assemblages of molecules. Winds and air currents generally are examples of mass movements of gas molecules. The draft of warm air ascending a chimney is another. All of the phenomena just listed, and many others, are due primarily to differences in the density of the gases in various parts of a system. Heavy gases are more strongly attracted by gravitational forces than light gases. Hence, as in a chimney, cold (relatively heavy) air, displace shot (relatively light) air, forcing the latter to rise. Similar phenomena on a grand scale are the principal cause of winds and air currents. They are representative of the physical process called convection. Such phenomena also occur in liquids. Mass movements of gases and liquids can also be caused in many other ways.

DIFFUSION PRESSURE

That diffusing gases sometimes result in the development of measurable pressures can be readily shown by certain simple experiments.

This apparatus consists essentially of a vertically arranged glass tube, to the upper end of which is attached, by means of rubber stopper, a cylindrically shaped porous clay cup. This cylinder is hollow and its thick walls are pierced by numerous minute capillaries. These pores are fine enough to prevent mass movement of gases at any appreciable rate, but the porous clay is not a membrane in the usual sense of the word. The lower end of the vertical glass tube dips in some coloured water. If such a porous clay cylinder, containing air, be enclosed within a bottle containing hydrogen gas a rapid bubbling of gas will occur from the lower end of the tube through the dye solution into which it dips. When the bottle is removed a rapid and sudden rise of liquid up the glass tube will ensue. This is followed by a slow subsidence in the level of the liquid in the glass tube, until it falls to that of the water in the beaker.

The explanation of this sequence of events is as follows. For reasons which are discussed later the rate of diffusion of hydrogen is greater than that of nitrogen or oxygen under comparable conditions. When the porous clay cup is first enclosed within the bottle of hydrogen, rapid diffusion of hydrogen gas occurs through the pores of the cup. This raises the total gas pressure within the cup, since outward diffusion of oxygen and nitrogen occurs at a much slower rate. Hydrogen gas diffuses into the cup in spite of the fact that this results in a greater total pressure inside the cup than in the surrounding atmosphere. The direction of the diffusion of the hydrogen gas is controlled entirely by its own differences in concentration, and is unaffected by the presence of other gases. The greater gas pressure inside the cup results in the outward bubbling of a mixture of all three gases at the lower end of the vertical glass tube. When the bottle is removed from around the porous clay cup, the direction of the diffusion of hydrogen is reversed, since the concentration of hydrogen gas is now greater inside of the cup. Since, while enclosed in the bottle of hydrogen, some of the nitrogen and oxygen diffused out of the cup and some was lost by the bubbling of gases from the lower end of the glass tube, this rapid outward diffusion of hydrogen results temporarily in a lower total gas pressure inside the cup than that originally present. Hence the solution rises rapidly in the glass tube. Finally, nitrogen and oxygen diffuse slowly inward, because of the reduced concentration of these gases inside of the cup, and the liquid slowly falls in the glass tube to its original level.

That the process of diffusion may result in the development of pressure also can be shown very strikingly by means of another simple experiment. If a pure rubber balloon containing only a little air be suspended in a closed bottle of carbon dioxide gas it will gradually become distended. Rubber membranes are quite readily permeable to the molecules of carbon dioxide but virtually impermeable to those of oxygen, nitrogen, or other gases of the atmosphere. Since, temperature remaining constant, the pressure exerted by any gas is directly proportional to its concentration (number of molecules per unit volume), it follows that the diffusion of the molecules of gases may be interpreted in terms of the differences in the partial pressure exerted by that gas in different parts of a system. Because of the greater partial pressure of carbon dioxide in the surrounding atmosphere, diffusion of this gas through the walls of the balloon continues until the partial pressure of the carbon dioxide is the same on the two sides of the rubber membrane. Since the initial partial pressure of carbon dioxide inside the balloon was virtually zero, and that in the bottle equivalent to one atmosphere, the inward diffusion of carbon dioxide gas results in a considerable distension of the balloon.

In the analysis of diffusion phenomena it is often clarifying to speak of the partial pressure of a gas as its diffusion pressure. Solutes and liquids may also be considered to possess a diffusion pressure, although the existence of such a physical quantity is less evident in such systems than in gases. Hence the concept that diffusion is the movement of the molecules of a substance from a region of its greater to a region of its lower diffusion pressure K often more satisfactory than the interpretation of diffusion phenomena in terms of concentration differences.

PRINCIPLE OF INDEPENDENT DIFFUSION

In the first experiment described in the preceding section it was shown that while one gas (hydrogen) was diffusing in an inward direction through the pores of a clay cylinder, other gases (oxygen and nitrogen) were simultaneously diffusing in an outward direction through the same pores. This exemplifies one of the most important principles governing diffusion phenomena, namely that the direction in which any substance will diffuse is controlled entirely by its own differences in diffusion pressure, and is not influenced by either the direction or rate of diffusion of other substances in the same system. Hence, in any given system, as for example, two adjacent plant cells, a number of substances may be diffusing in one direction across the intervening membranes, while simultaneously other compounds may be diffusing in the opposite direction across the same membranes. Each one of these individual substances will diffuse in the direction determined by its own differences in diffusion pressure, and at a speed which is determined by the factors which are influencing the diffusion of that particular substance.

FACTORS INFLUENCING THE RATE OF DIFFUSION OF GASES

DENSITY OF THE GAS

Different gases diffuse at different rates even when influenced by the same set of environmental factors. Hydrogen, for example, diffuses more rapidly than any other gas. The same Thomas Graham who conducted some of the earliest studies on the properties of colloidal systems also was one of the first investigators to study quantitatively the phenomenon of gaseous diffusion. He discovered the principle, often called Grahams Law of Diffusion, that the relative speeds of diffusion of different gases are inversely proportional to the square roots of their relative densities. By relative density is meant the weight of a given volume of gas as compared with the weight of the same volume of hydrogen. The relative density of oxygen is 16.

Hydrogen gas will therefore diffuse four times as rapidly as oxygen gas under the same conditions of temperature and pressure. The relative density of carbon dioxide is 22; hence by similar reasoning it is apparent that hydrogen gas will diffuse nearly five times as rapidly as carbondioxide gas.

1 Since a molar weight of any gas occupies a volume of 22.4 liters at standard conditions, molar weights of gases are in themselves a measure of the relative density of gases. Since the molar weight of hydrogen (H_2) is 2.016, the relative density of any gas on the basis H i, is equal to its molar weight divided by 2.016 (usually rounded off to 2).

TERNPERATURE

Increase in temperature increases the speed of diffusion.This is due, at least in part, to the correlated increase in the kinetic activity of the molecules of the diffusing substance. Actual measurements of the temperature coefficient 2 of diffusion generally yield values between 1.2 and 1.3. Such values are characteristic of any purely physical process such as diffusion.

DIFFUSION GRADIENT

The fact that the direction of the diffusion of gases is from a region of their greater diffusion pressure to a region of their lesser diffusion pressure has already been emphasized. The speed of diffusion is also influenced by differences in diffusion pressure. In general the greater the difference in diffusion pressures between the two regions, the more rapidly diffusion will occur. The rate of diffusion is influenced, however, not only by the difference in diffusion pressures but also by the distance through which the diffusing molecules must travel. These two factors are components of what may be called the diffusion pressure gradient or concentration gradient. In both parts, A and B of this diagram, oxygen is represented as diffusing from a region in which its diffusion pressure is maintained at 760 mm.Hg (atmospheric pressure) into a region in which its diffusion pressure is just half as great. The length of the connecting tube is, however just twice as great in A as in B. Under the conditions postulated the rate of diffusion from the region of greater diffusion pressure to the other will be just twice as rapid in B as in A. The diffusion pressure gradient is equivalent to the difference in the diffusion pressures between the delivering and receiving ends of the diffusion system divided by the length of the distance between.

The greater or steeper this gradient the more rapidly diffusion will occur. The steeper the gradient, the more rapid the change in diffusion pressure per unit of length along the axis of the diffusion gradient. The steepness Of a diffusion pressure gradient may be changed by varying either factor, the difference in diffusion pressures, or the length of the gradient.

The temperature coefficient (Q10) of any process, physical, chemical, or physiological, is defined as the number of times that the rate of the process increases with a 200C. rise in temperature. If the rate of the process is doubled the temperature coefficient is 2, etc.

CONCENTRATION OF THE MEDIURN THROUGH WHICH DIFFUSION OCCURS

In general the more concentrated the medium, i.e. the more molecules per unit volume in the medium through which the diffusing molecules must pass, the slower the rate of diffusion. Bromine gas diffuses more rapidly through a vacuum than through air.

DIFFUSION OF SOLUTES

The molecules or ions of a solute possess sufficient kinetic energy to move from place to place within the limits of a solution. The simplest method of demonstrating the diffusion of a solute is to introduce a crystal of copper sulphate, or some other compound which is coloured when in solution, into the bottom of a tall glass cylinder filled with water. The cylinder should then be placed in an environment of equable temperature where it will be free from disturbance. The diffusion of the molecules or the ions which pass into solution in the water can be followed by the slow change in colour of the water. One of the most striking facts illustrated in such experiments is the extremely slow rate of diffusion of solutes through water. This is partly due to the fact that in such an experiment the steepness of the diffusion gradient decreases with time, but principally to the fact that the densely packed molecules of the liquid enormously impede the diffusion of the dissolved molecules or ions.

The true rate of diffusion of solutes is probably even less than the rates indicated in such experiments, since in all such setups convection currents may develop in the water and aid in the distribution of the solute throughout the body of the liquid. The diffusion of solutes is often demonstrated by employing a gel rather than a liquid as the medium into which diffusion occurs. Such a technique avoids errors introduced by the development of convection currents.

The direction of the diffusion of any solute occurs in accordance with its own differences in diffusion pressure, regardless of the rate or direction of diffusion of other solutes in the same system. The rate of diffusion of solute particles is governed by principles essentially similar to those which control the rate of diffusion of gases and is controlled by the following factors:

SIZE AND MASS OF THE DIFFUSING PARTICLE

Small molecules or ions diffuse more rapidly than large ones. A hydrogen ion, for example, diffuses many times more rapidly than a glucose molecule. Similarly, highly hydrated ions diffuse more slowly than those which have fewer water molecules bound to them, since the association of water of hydration with a molecule or ion in effect increases its size. The mass of the particle will also be a factor influencing the speed of its diffusion. As between two particles of the same size, but different masses, the heavier particle will diffuse more slowly.

TEMPERATURE

The kinetic activity, and hence the rate of diffusion of solute molecules, increases with increase in temperature.

DIFFUSION GRADIENT

The steeper the diffusion gradient, the more rapidly solute particles diffuse.

SOLUBILITY

In general, the more soluble a substance is in a liquid the more rapidly it will diffuse through that liquid. This influence of solubility upon diffusion rates can be interpreted principally in terms of its effect upon the diffusion gradient, since obviously steeper gradients can be built up if the solute is very soluble in the liquid than if it is only slightly soluble.

Chapter 8

Osmosis and Osmotic Pressure

Although the freedom of movement of the molecules of liquids is to some extent restrained by internal cohesive forces, they also possess kinetic activity. Like gases and solutes, therefore, liquids exhibit diffusion phenomena. If, for example, water is brought into contact with another liquid such as ether with which it is only slightly miscible, a slow diffusion of water molecules into the ether will occur. Simultaneously a slow diffusion of the molecules of the ether will take place into the water. Such diffusion will continue until the two liquids are mutually saturated.

OSMOSIS

This is by far the most familiar process involving the diffusion of liquids. An understanding of the dynamics of this process, and of the significance of the physical quantity termed osmotic pressure is essential to an interpretation of the water relations of plant cells and tissues.

Let us first consider a sac-like membrane of collodion is nearly filled with a strong sucrose solution and immersed in a beaker of water. The top of the sac is tightly plugged with a rubber stopper. The collodion membrane is so prepared that it is permeable to water, but impermeable or practically so to sucrose; in other words it is differentially permeable. It is also slightly elastic.

After a short time the originally limpsac becomes rigidly distended. This is due to the movement of water through the collodion membrane into the interior of the sac. Such a diffusion of water is an as a result of the entrance of water is exerted against the inside wall of the membrane. This pressure is counterbalanced by an equal and oppositely directed pressure exerted upon the solution by the distended wall. If the membrane is virtually impermeable to the solute eventually the entire system will come to equilibrium after which there is no further increase in the volume of water inside of the membrane. When the movement of the solvent is referred to in discussing osmosis it is always the net movement which is meant. Solvent molecules will always be moving across the membrane in both directions, but except when an equilibrium has been attained, more molecules will move per unit of time in one direction than in the other. It is the presence of this differentially permeable membrane which gives to osmosis its distinctive aspect as compared with other diffusion processes. If the membrane were permeable to the solute particles, they would diffuse in one direction, while the molecules of the solvent were diffusing in the other; this would be another example of the principle of independent diffusion.

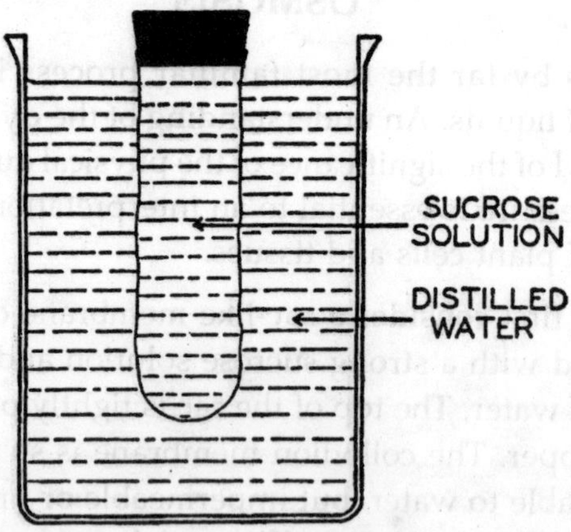

Fig. Apparatus for the demonstration of osmosis through a collodion membrane

TERMINOLOGY

The conditions under which osmosis can occur are so varied that it is impossible to characterize the process in terms of a simple definition. The following statement will, however, apply to most cases:

Osmosis is the diffusion of a liquid through a differentially permeable membrane into a solution in which the solvent is that same liquid or into another liquid with which it is miscible. In living organisms water is the only important liquid which moves from cell to cell by osmosis, hence the further discussion of this process will be in terms of water and aqueous solutions. The attainment of a satisfactory comprehension of osmosis and osmotic pressure has been delayed both by confusion in theoretical interpretations as well as by lack of a consistent terminology. Many obscurities have resulted from the application of the term osmosis to the movement of solutes through membranes. The use of the term in this sense has not only led to many misunderstandings, hut also to positive errors and misinterpretations of data.The movement of solutes through a membrane is a diffusion process; it maybe simple diffusion, or it may be complicated by a number of factors which do not enter the phenomenon of simple diffusion. Both from the historical and theoretical standpoint it seems desirable that the term osmosis he restricted to the sense in which it is used in this discussion.

Osmotic pressure may be defined as the maximum pressure which can be developed in a solution when separated from pure water by a rigid membrane permeable only to water. In order to determine the osmotic pressure of a solution it would be necessary to enclose it in a rigid membrane permeable only to water, to immerse this membrane in pure water, and to exert just enough pressure on the solution (by means of a leak-proof piston, for example) to prevent any increase in its volume due to the entrance of water.The osmotic pressure of the solution is quantitatively equal to the imposed pressure. If the latter is 15 atmos., then the osmotic pressure of the solutionis 15 atmos., etc.

It is a common practice to speak of solutions as possessing an osmotic pressure whether they are under such conditions that a pressure can develop within them or not. A weight molar solution of sucrose contained in a bottle may therefore be spoken of as possessing an osmotic pressure of 27 atmos at 25C. In other words the term osmotic pressure is commonly used to denote the potential maximum pressure which would develop in a solution were it placed under the necessary conditions. Its use will be confined to this sense in this book.The term osmotic pressure is also often used as a designation for the actual pressures developed as a result of osmosis, a procedure which leads to confusion. Actual pressures developed during osmosis are seldom equal to the osmotic pressure as defined above. If a solution having an osmotic pressure of 12 atmos. be enclosed in a stoppered membrane permeable only to water which is in turn immersed in a solution having an osmotic pressure of 8 atmos., water will diffuse inwards until at equilibrium the actual pressure developed in the internal solution will be (disregarding any minor effect due to dilution) 4 atmos. Its osmotic pressure will still be nearly 12 atmos, however. Even if the external liquid is pure water. The actual pressure developed in the internal solution would not be equal to its original osmotic pressure unless the membrane is completely inelastic. Inward diffusion of water results in a dilution of the solution which, as we will see later, always results in a decrease in its osmotic pressure. To avoid the inevitable confusion caused by use of the same term in a dual sense we might distinguish between the potential (i.e. maximum) osmotic pressure and the actual osmotic pressure of a solution. In this discussion however, we shall employ the term turgor pressure to refer to the actual pressure developed as a result of osmosis. This term has long been employed as a name for the actual pressures developed in plant cells. Since these are largely of osmotic origin the name turgor pressure seems equally appropriate for physical systems. Actually the turgor pressure developed in a solution enclosed within a membrane may be increasing while its osmotic pressure is decreasing, as shown in the last example mentioned above. This indicates clearly that each of these terms osmotic pressure and turgor pressure refers to a different physical quantity. In an osmometer set up as just described the maximum turgor pressure which could be developed will never exceed the osmotic pressure.

1. Osmotic pressure is sometimes defined as the pressure which must he imposed on a solution to prevent any increase in its volume under the conditions specified above, but this usage is particularly unsatisfactory in the interpretation of biological phenomena. While the magnitudes of such an imposed pressure and the osmotic pressure are mathematically equal, physically they are two entirely different quantities.
2. The osmotic pressure of a solution has also been variously called the osmotic potential, osmotic power, osmotic value, and osmotic concentration of a solution. Such terms are frequently encountered in the literature of plant physiology.

THE MECHANISM OF OSMOSIS

Like a gas, water or any other liquid maybe considered to possess a diffusion pressure. The only two factors which influence the diffusion pressure of a pure liquid are pressure and temperature. The imposition of pressure from an external source will raise the diffusion pressure of a liquid. For example, if a pressure of 10 atmos. be exerted upon water in a closed vessel by means of a leak-proof piston, the diffusion pressure of that water will be increased by 10 atmos. Under certain conditions negative pressures or tensions may develop in liquids. These result in a reduction in the diffusion pressure of the liquid in an amount equal to the magnitude of the tension.

Although the diffusion pressure of water is also influenced by temperature no interpretation of the influence of this factor will be undertaken. In order to simplify the following discussion it will be consistently assumed that all parts of every osmotic system considered are at the same temperature. However, when a substance is dissolved in water the diffusion pressure of the water in the resulting solution is decreased as compared with that of pure water at the same temperature and pressure. This diminution in diffusion pressure is proportional, within a wide range of solution concentrations, to the number of solute particles present in a given volume (i.e. a given number of molecules) of the solvent. In a solution, therefore, the diffusion pressure of the solvent may be influenced by the three factors of

1. Temperature,
2. Pressure, and
3. The ratio of solute particles to solvent molecules.

When a solution is confined within a membrane permeable only to water and that membrane is immersed in water there will be a net movement of water through the membrane into the solution, because of the excess diffusion pressure of the water on the water side of the membrane. The passage of water through the membrane results in the development of a turgor pressure on the solution side of the membrane. Since the maximum possible pressure which can develop in the solution is equal to the excess of the diffusion pressure of pure water over the diffusion pressure of the water in the solution, the following relation holds, providing all parts of the system are at the same temperature and under the same external pressure:

Osmotic pressure Diffusion pressure of pure water Diffusion pressure of water in the solution.

In other words the osmotic pressure of a solution is a measure of what may be called the diffusion pressure deficit of the water in that solution. By this term, as applied to any given solution, is meant the amount (usually expressed in atmospheres) by which the diffusion pressure of the solvent in that solution is less than the diffusion pressure of the pure solvent when the latter is at the same temperature and under atmospheric pressure. For example, if a solution has an osmotic pressure of 10 atmos., this signifies that the diffusion pressure of the water in that solution is just 10 atmos. less than the diffusion pressure of pure water under atmospheric pressure at the same temperature; in other words its diffusion pressure deficit is 10 atmos. The osmotic pressure of a solution is a measure of the diffusion pressure deficit of the water in that solution only when the solution is not under a pressure or a tension (except, of course, atmospheric pressure) since these factors also affect the diffusion pressure of water. In plant cells, as we shall see later pressures and tensions must constantly be taken into account in evaluating the diffusion pressure deficit of the water in the cell sap.

When two aqueous solutions, both initially subjected only to atmospheric pressure, are separated by a membrane permeable only to water, diffusion of water will take place towards the solution of greater osmotic pressure. The diffusion pressure deficit of the internal solution A is 20 atmos.; that of the external solution B, 12 atmos. Water always moves in any osmotic system towards the region of its lesser diffusion pressure or in other words towards the region of its greater diffusion pressure deficit. In this particular example, disregarding dilution, the internal solution will exert a turgor pressure of 8 atmos. at equilibrium. Correspondingly the membrane will exert a wall pressure of 8 atmos. against the enclosed solution.

Since subjection to a pressure raises the diffusion pressure of water by the amount of the imposed pressure, the diffusion pressure deficit of the water in the internal solution at equilibrium will not be 20 atmos., but 12 atmos. In other words, at equilibrium, the diffusion pressure deficit of the water will be equal on both sides of the membrane. In the preceding discussion it has been tacitly assumed that water diffuses through the membrane during osmosis in the liquid state. Many investigators consider, however, that actual diffusion of water across the membrane occurs in the form of water-vapour. Callendar who first proposed such a vapour pressure theory of osmosis, assumed the existence in differentially permeable membranes of minute capillary pores too small to be wetted by liquids, but freely permeable to molecules in the vapour state. The essential concepts in the vapour pressure theory of osmosis can be presented in terms of two simple experiments.

The presence of the sucrose molecules decreases both the diffusion pressure and the vapour pressure of the water in the solution. In fact, increase in the diffusion pressure deficit (osmotic pressure) of a solvent caused by the introduction of a solute is accompanied by a proportionate decrease in its vapour pressure. As a result the vapour pressure just above the layer of pure water will be greater than that above the layer of the solution, and a gradient of vapour pressures will be established, decreasing in magnitude from the water to the solution. Water will therefore move from the region of high vapour pressure to the region of low vapour pressure, and the solution will steadily gain in volume at the expense of the pure water. Ultimately all of the water will be transferred to the sucrose solution. The transfer of water from one arm to the other has resulted from the difference in vapour pressures of the two liquids and is therefore directly correlated with their differences in diffusion pressures.

There will be an increase in the volume of the solution in x, and a decrease in the volume of the water in y. If we assume that the water moves through the membrane in the form of vapour there is no essential difference between this phenomenon and the one first described.In the foregoing discussion the effect of temperature on the osmotic movement of water has been disregarded. If different parts of an osmotic system are maintained at different temperatures this will have an effect on the osmotic movement of water. However unless the difference in temperature on the two sides of the membrane is very considerable the effects of this factor onosmosis are slight, and usually can be disregarded.

The attainment of an osmotic equilibrium between two solutions, or between a solution and water does not necessarily imply an equalization of solute concentration, nor of pressure, nor of temperature. It does imply however, an equalization of diffusion pressures and an examination of the circumstances is necessary to determine the exact mechanism by which they have been brought into equilibrium in any given system.

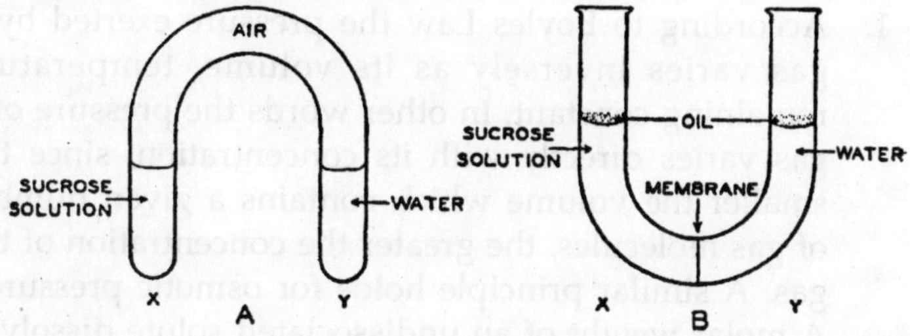

Fig. Diagrams illustrating discussion of the vapour pressure theory of osmosis

ANALOGIES BETWEEN OSMOTIC PRESSURE AND GAS PRESSURE

The important fact is that while the osmotic pressures of weight molar solutions of non-electrolytes approximate the same value, which is in the neighbourhood of vant Hoffs theoretical value of 22.4 atmospheres, the values for weight molar solutions of electrolytes are much higher.

Vant Hoff showed that a remarkable analogy exists between the osmotic pressures of dilute solutions and the pressures exerted by gases. Vant Hoff pointed out that the osmotic pressure of a dilute solution is equal to the pressure that the solute alone would exert were it present as a gas at the same temperature in the same volume as that occupied by the solution. A dilute solution was considered by vant Hoff to be one in which the volume of solute was small in proportion to the volume of solvent. For example, if we could imagine the complete and instantaneous removal of all of the water in a sucrose solution without any change in the spatial relations of the solute molecules, the residual sucrose molecules would be in a gaseous state. According to vant Hoff the osmotic pressure of this solution would be equal to the gas pressure which would be exerted by such a hypothetical sucrose gas.

Furthermore, he showed that the osmotic pressures of dilute solutions obey laws analogous to those which describe the relations of gases to variations in temperature volume and pressure. The analogies will now be briefly summarized:

1. According to Boyles Law the pressure exerted by a gas varies inversely as its volume, temperature remaining constant. In other words the pressure of a gas varies directly with its concentration, since the smaller the volume which contains a given number of gas molecules, the greater the concentration of the gas. A similar principle holds for osmotic pressures. A molar weight of an undissociated solute dissolved in 1000 g. of water.
2. According to the law of Gay-Lussac the pressure exerted by a gasaries directly with the absolute temperature, if the volume of the gas remains constant. The osmotic pressure of a solution is therefore proportional to the absolute temperature.
3. According to Avogadros Hypothesis equal volumes of all gases under identical conditions of temperature and pressure contain equal numbers of molecules and hence exert equal pressures. One mol of any gas at standard conditions Ċ.; 1 atmos. pressure) occupie 4 liters. If this gas be compressed to a volume of 1 liter, it will exert a pressure of 22.4 atmos. (Boyles Law). Similarly one mol of any undissociated solute dissolved in a liter of water exerts an osmotic pressure of approximately 22.4 atmos. at 00 C. Actually the analogies between gas pressures and osmotic pressures as discussed by vant Hoff hold with a fair degree of approximation for solutions up to concentrations of at least one molar.

This is the same pressure that it would exert were it a gas confined in the same volume under standard conditions. It follows therefore that equal osmotic pressures of solutions of undissociated solutes indicate equal concentrations of solute molecules and that Avogadros Hypothesis may be applied to osmotic pressures as well as to gas pressures. In other words the theoretical osmotic pressure of a weight molar solution of any undissociated solute is 22.4 atmos. at 00 C.

Electrolytes, as first shown by De Yries in 1884 exhibit greater osmotic pressures than their molar concentrations would lead one to expect. This fact remained unexplained until the formulation of the ionization theory by Arrhenius in 1887. In fact DeVries data were used by Arrhenius as strong evidence for his theory. Apparently each ion produced by the dissociation of an electrolyte is just as effective osmotically as an undissociated molecule. Hence all electrolytes produce higher osmotic pressures than the theoretical values for undissociated compounds; the exact magnitude of the pressures produced will depend upon their degree of dissociation. For example, a molar solution of NaCl is about 75 per cent dissociated. This means that 75 out of every 100 molecules of NaCl present have dissociated into ions. There will therefore be 1.75 times as many particles (ions and molecules) present as in a molar solution of a non-electrolyte. The osmotic pressure of a molar solution of NaCl would therefore be theoretically 22.4 × 1.75 39.30 atmos.

The hydration of molecules or ions also influences the osmotic pressure of a solution. Water molecules which are bound to solute particles as water of hydration are no longer effective as a part of the solvent. In effect a solution containing hydrated solute particles is therefore more concentrated than its molarity would indicate. This accounts, at least in part, for the fact that the experimentally determined values of osmotic pressures are often in excess of the theoretical ones which would be expected according to the Vant Hoff analogy with the gas laws. For example, a molar solution of sucrose should theoretically have an osmotic pressure of 22.4 atmos, at o0 C.Actually its measured pressure is found to be 24.83 atmos.This discrepancy is believed to be due to the hydration of the sucrose molecules. One sucrose molecule binds six molecules of water of hydration.

Hence one mol of sucrose will bind six mols of water. One thousand grams of water is equivalent to 55.5 mols of water.

Since the water of hydration is not free to act as a solvent, in effect the solution consists of 1 mol of sucrose dissolved in 49.5 mols of water. A simple calculation (22.4 × 55.5 49.5) yields the value 25.12 atmos. as the theoretical osmotic pressure of a weight molar sucrose solution if hydration of the sucrose molecules is taken into account. This value is in close agreement with those which have been obtained experimentally.

Vant Hoffs demonstration of the analogy between osmotic pressure and gas pressure led many subsequent investigators to consider that osmotic pressure results from the impacts of the solute molecules against the interior wall of the membrane, an assumption which vant Hoff himself was careful to avoid. One reason for the persistence of this hypothesis of osmotic pressure has been the difficulty in understanding why if the osmotic pressure is due to the solvent molecules it should be approximately equal to the pressure which would be developed if the solute molecules were present in the same volume in the form of a gas.

The solvent pressure hypothesis holds that the osmotic pressure which develops in an osmometer or similar closed system is due to the excess of the inwardly directed diffusion pressure of the solvent over the outwardly directed diffusion pressure of the solvent. Haldane has shown that this excess inwardly directed diffusion pressure of the solvent is mathematically equal to the pressure which the solute molecules inside the osmometer would exert if present as a gas in the same volume as the solution. Thus it has become possible to reconcile the concept that osmotic pressure is due to the solvent molecules with vant Hoffs classical discoveries regarding the analogy between osmotic and gas pressures.

THE MEASUREMENT OF OSMOTIC PRESSURES

The osmotic pressure of a solution, as such, can be measured only by the direct manometric method. Precise results can be obtained by this method only at the cost of an elaborate technique and infinite precautions. Hence the actual number of measurements of osmotic pressures which have been made by this direct method are very limited. As is well known there is a direct proportionality between the osmotic pressure, lowering of the vapour pressure, elevation of the boiling point, and depression of the freezing point of solutions. Hence osmotic pressures of solutions can be calculated from the determined results of one of these three physical quantities. Most frequently determinations are made of the depression of the freezing point of a solution, and its osmotic pressure calculated from this. Since the theoretical freezing point depression of a weight-molar solution of an unionized substance is 1.860 C., and the theoretical osmotic pressure of such a solution is 22.4 atmos., an equation relating freezing point depressions and osmotic pressures is easily derived. If we let A represent the freezing point depression of a solution, its osmotic pressure (O.P.) may be calculated as follows:O.P.: 22.4 = A: 1.861.86O.P. = 22.4 A22.4 A1.86O.P. = 12.04 A4

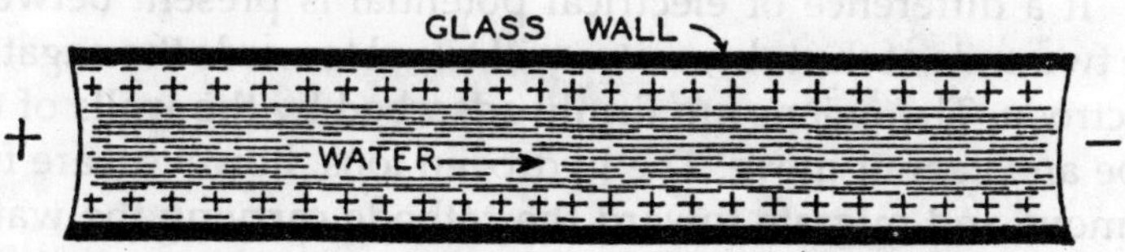

Fig. Diagram to illustrate electro-osmosis

For example, a solution for which the determined freezing point depression is 0.930 would have an osmotic pressure of 11.10 atmos. (12.04 × 0.930).This equation appears to be approximately correct over a wide range of concentrations since deviations from the theoretical osmotic pressures of solutions are accompanied by almost strictly proportional deviations from their theoretical freezing point depression values. This method is often called the cryoscopic method of determining osmotic pressures. The osmotic pressures as determined by this method are as at the freezing point temperature of the solution.

ELECTRO-OSMOSIS

Under certain conditions pure water will diffuse from one side of a membrane to the other under the influence of a difference of electrical potential, a process which is called electro-osmosis. When water is confined in a capillary glass tube adsorption of OH (or HCO_3) ions imparts a negative charge to the walls of the tube. Adjoining them is a layer of H_1 ions equal in number to the adsorbed anions. In other words ions become distributed as an electrical double layer just as they do around a colloidal particle.

If a difference of electrical potential is present between the two ends of the tube, water will travel towards the negative electrode. The anions are firmly adsorbed by the walls of the tube and cannot move. The hydrogen ions, however, are free to move and migrate toward the cathode carrying the water, of which they are a part, along with them. In other words the moving A slightly more accurate form of this equation has been derived by Lewis. The tables of Harris and Gortner based on the Lewis equation, are very useful for converting freezing point depressions into osmotic pressures. Cations in the water glide past the stationary anions which adhere to the walls of the tube. Since many membranes are essentially porous in structure,conditions similar to those just described may exist in each capillary of the membrane, and electro-osmosis may occur through such membranes in essentially the same way that it occurs through tubes of small bore. Since differences of electrical potential are frequently present in living organisms it seems entirely probable that electro-osmotic flow of water occurs in living tissues.

Chapter 9

Imbibition

If a handful of the seeds of any species which do not possess seed coats impermeable to water be dropped into a vessel of water, within a few hours they will have swollen visibly. Many other materials will swell in a similar way when immersed in water. Among these are starch, cellulose, agar, gelatin and kelp stipe. Some substances will swell similarly when immersed in other liquids. All of these phenomena are examples of the process usually called imbibition. The amount of water which may enter substances in imbibition is often very great in proportion to the dry weight of the substance which swells. A piece of dried kelp stipe, for example, can absorb as much as fifteen times its own weight of water.

Mater may be imbibed as a vapour as well as in the liquid state. The swelling of doors and wood work generally during damp weather is a familiar example of this phenomenon. Plant structures, if sufficiently low in water content, also imbibe water-vapour. The water content of air-dry seeds, for example, generally fluctuates with the vapour pressure of the atmosphere, rising with an increase in vapour pressure, and vice versa.

THE DYNAMICS OF IMBIBITION

Imbibition is usually considered to be basically a diffusion process, but capillary phenomena are probably also involved. Imbibing substances are often permeated with minute submicroscopic capillaries, and it is impossible to determine how much of the liquid enters by diffusion and how much by capillary movement through invisible pores. Fundamentally, however, the cause of imbibition may be regarded as a difference in the diffusion pressure between the liquid in the external medium and the liquid in the imbibant. As long as the latter is less than the former movement of water into the imbibing substance will continue. An equilibrium will be reached, as in diffusion or osmotic phenomena, only when the diffusion pressure of the water in the two parts of the system has attained the same value.

A substance which imbibes water does not necessarily imbibe other liquids. Dry kelp stipe, for example, swells enormously when immersed in water but does not swell when immersed in ether or other organic liquids. Contrariwise, a piece of rubber does not imbibe water, but does imbibe appreciable amounts of ether and other organic compounds when in contact with themin either the liquid or the vapour state. Obviously a difference in diion pressures between the liquid in an imbibant and in its surrounding medium is not the only requisite for the occurrence of imbibition. Certain specific attractive forces between the molecules of the imbibant and the imbibed liquid must also be present. In default of such specific attractions imbibition fails to occur, even if all other necessary conditions for the process are fulfilled. While perhaps not an invariable principle it appears to be generally true that only strongly polar substances imbibe strongly polar liquids such as water, while non-polar substances such as rubber imbibe only non-polar or very weakly polar organic liquids.

Since in living organisms water is the only liquid imbibed, the further discussion will disregard other types of imbibitional phenomena. The physical relationship between the imbibed water and the imbibant is undoubtedly a complex one. The bulk of the imbibed water in any system is probably adsorbed on the molecules or micelles which constitute the structural units of the imbibing material. The possible relation of adsorbed water to the adsorbing particles in colloidal systems have already been considered.

There is good reason for believing that similar physical relations exist between the imbibed water and unit particles in an imbibant. Such water may be pictured as either actually in solution in the adsorbing particles, or as adsorbed as a shell of from one to many molecular layers in thickness on their surface. It is possible that either or both of these conditions obtain in many systems containing imbibed water, but the second of these pictures appears to be more likely. It is also probable that a portion of the imbibed water enters and occupies minute submicroscopic capillaries which ramify between the component molecules or micelles of the imbibant. The result of imbibitionis often the production of a system which in its essential properties may be regarded as a gel. The physical status of the water in a system into which it has been imbibed and in an elastic gel is undoubtedly very similar. In fact the imbibition of a dispersion medium by a dry substance is generally regarded as a method of gel formation.

VOLUME CHANGES IN IMBIBITION

The volume of the imbibant always increases during imbibition. The final volume of the entire system (liquid+ imbibant) is always less, however, than the sum of the initial volume of the liquid and the imbibing substance. In other words a contraction in the volume of the system, liquid + imbibant, occurs during the process of imbibition. This volume shrinkage is not principally due to the occupancy of minute spaces within the substance of the imbibing material by the liquid, as might be surmised, although in some systems a part of the contraction in volume possibly may be accounted for in this way. The explanation is undoubtedly to be ascribed to the fact that the adsorbed water molecules are definitely oriented in relation to the adsorbing surfaces, and hence occupy less space than when in the free state. This is equivalent to a compression of the adsorbed water so that its density is greater than that of free water. Energy Relations of Imbibition. The process of imbibition always results in the release of heat. Liberation of heat during imbibition may be easily detected by allowing dry starch or some other material with a high imbibitional capacity to imbibe water while contained in a calorimeter, and noting the change in temperature.

The adsorption of water molecules which occurs when they are imbibed results in a loss of a large part of their kinetic energy, which reappears in the system as heat energy. This increase in the temperature of the system corresponds to an increase in the kinetic activity of all except the adsorbed molecules. The essential energy change in the process of imbibition is a loss of kinetic energy by the adsorbed molecules, and its transference to the other molecules of the system. Their increased kinetic activity is the cause of the observed rise in temperature. The greatest evolution of heat accompanies the initial stages of imbibition. This is to be expected, as a larger proportion of the first molecules of water imbibed would be firmly adsorbed than of those which enter the imbibant later.

FACTORS INFLUENCING IMBIBITION

In discussions of the factors influencing imbibition it is often necessary to distinguish between the influence of a factor upon the rate of imbibition, and upon the total amount of water which will be held in the system when an equilibrium has been attained. The principal factors which influence the entrance of water into an imbibant are as follows:

TEMPERATURE

The rate of imbibition increases with increase in temperature. Equilibrium will therefore be attained more slowly at lower temperatures, but the amount of water held in an imbibant at equilibrium will be slightly greater at lower than at higher temperatures. This is in accord with the principles of the influence of temperature upon adsorption processes generally.

OSMOTIC EFFECTS

Water moves by imbibition into a substance only when its diffusion pressure is in excess of the diffusion pressure of the water in the imbibant. The introduction of a solute into water invariably has the effect, as already discussed, of reducing the diffusion pressure of the water in that solution as compared with pure water. The osmotic pressure of a solution is a measure of its diffusion pressure deficit. Hence the magnitude of the osmotic pressure of the surrounding solution would be expected to influence both the rate of imbibition and the equilibrium water content of an imbibant.

One of the most comprehensive studies dealing with osmotic effects in relation to imbibition has been conducted by Shull who used the seeds of the cocklebur (Xantlziurn pennsylvanicum) as the imbibing material. Samples of the seeds of this species were allowed to come to equilibrium with solutions of sodium chloride and lithium chloride of different osmotic pressures, after which the amount of water which had been imbibed was determined. The coats of these seeds are permeable to water, but virtually impermeable to these solutes. Were it not for this latter fact complicating effects due to the specific influences of the ions employed, as described in the next section, might invalidate the results of experiments of this type.

The general principle shown by the results of these experiments is that with increase in the osmotic pressure of the solution in which the seeds are immersed, the amount of water held by imbibition per unit of dry weight at the equilibrium point decreases. Since at equilibrium the diffusion pressures of the water in the imbibing substance and in the surrounding liquid must be equal the basis for this osmotic effect upon imbibition is evident. If the diffusion pressure in the solution is relatively low (high osmotic pressure) the diffusion of less water into the imbibing substance is necessary to raise the diffusion pressure of the water in it to a value equal to the diffusion pressure of the water in the circum ambient liquid, than if the diffusion pressure in the surrounding liquid is high. The relation between the osmotic pressure and the amount of water imbibed is not, however, a strictly proportional one. At the lower end of the range of concentrations employed, an increase of a few atmospheres in osmotic pressure causes a decrease in the amount of water imbibed, in terms of the air-dry weight of the seeds, of nearly15 per cent. Near the upper end of this range of concentrations an increase in osmotic pressure of several hundred atmospheres is required to produce an equivalent change in the volume of water imbibed. This is further evidence that the first increments of water passing into any imbibant are held by tremendously greater forces than those which are imbibed subsequently.

IONIC EFFECTS

The amount of water which will be imbibed from a given solution may also be influenced by the species of ions in that solution. If agar or kelp stipe, for example, is allowed to come to equilibrium with equimolar solutions of lithium, sodium, ammonium, and potassium chlorides swelling will not be equal in the different solutions. The greatest imbibition will occur in lithium chloride solution, next greatest in sodium chloride, next greatest in ammonium chloride, and least in the potassium chloride solution. In all of the solutions swelling will be less than in pure water.

At first sight it would seem that this might be ascribed to an osmotic effect.The potassium chloride solution might be assumed to have a higher osmotic pressure than the lithium chloride solution, due to differences in dissociation or hydration, and hence to exert more of a retarding effect upon imbibition than the other solutions. Actually the opposite condition holds, hence this influence of electrolytes upon imbibition cannot be explained in terms of an osmotic effect. Apparently, since all of these electrolytes give rise to the same anion the cause of this phenomenon must be looked for in terms of specific effects of the cations upon swelling.

Many other processes and physical properties are affected by this and similar ionic sequences. Such ionic series as these are known as lyotropic series, and such effects of ions as are found to be arranged in sequences of this sort are called ljo tropic effects.

Sometimes the order of the effect of a lyotropic series is from left to right and sometimes the reverse depending upon the property affected. As arranged above the series of univalent cations represents (reading from left to right) the order of their increasing effectiveness in checking the imbibition of water by kelp stipe. The same is true of the series of bivalent cations.The anions in the third series do not differ very greatly in their effects upon the swelling of kelp stipe, but are known to influence many other processes the magnitude of their effects being in the order given.

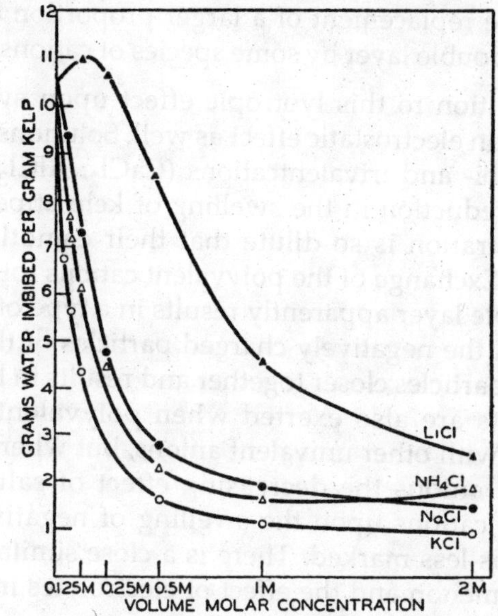

Fig. Influence of cations on imbibition of water by kelp stipe

These so-called lyotropic effects are apparently due largely to differences in the degree of hydration of various kinds of ions although other factors are probably also involved. Their lyotropic effect on the swelling of kelp stipe, agar, and other similar materials which bear a negative charge with respect to water probably may be explained as follows:

The negative charge on the kelp stipe is apparently due to adsorption of hydroxyl ions, leaving hydrogen ions in the outer zone of the electrical double layer. When immersed in a solution of an electrolyte not only is water imbibed, but an exchange of cations occurs, some of the cations in the electrolyte replacing hydrogen ions in the kelp stipe. The displaced hydrogen ions of the double layer pair off with the residual anions of the electrolyte. The water molecules associated with the cations which enter the double layer will contribute to the total hydration of the kelp. The different degrees of swelling caused by the different cations is probably due partly to the different numbers of water molecules associated with individual cations of various species, and partly to the replacement of a larger proportion of hydrogen ions in the double layer by some species of cations than others.

In addition to this lyotropic effect upon swelling, ions often exert an electrostatic effect as well. Solutions of chlorides producing bi- and trivalentcations ($CaCl_2$, $AlCl_3$, etc.) result in a great reduction in the swelling of kelp stipe even when the concentration is so dilute that their osmotic effects are negligible. Exchange of the polyvalent cations for the H + ions in the double layer apparently results in a type of flocculating effect upon the negatively charged particles of the kelp. This brings the particles closer together and results in less swelling. Such effects are also exerted when polyvalent cations are associated with other univalent anions, but when paired with polyvalent anions the decreasing effect of salts producing polyvalent cations upon the swelling of negatively charged imbibants is less marked. There is a close similarity between this phenomenon and the effect of electrolytes in flocculating negatively charged sols.

EFFECTS OF H-ION CONCENTRATION

Hydrogen ions in any appreciable concentration affect the swelling of a negatively charged imbibant adversely while hydroxyl ions in any appreciable concentration cause greater swelling of such systems than occurs in pure water. Exactly the reverse effects are exerted by hydrogen and hydroxyl ions upon the swelling of positively charged imbibants. The swelling of amphoteric gel-forming substances such as gelatin and certain other proteins is influenced in an even more complex manner by the pH of the swelling medium.

Minimum swelling occurs at the isoelectric point, and with increase or decrease of pH from this point increased swelling occurs. At relatively low or relatively high pH values a decreasing effect upon swelling is again evident.

IMBIBITION PRESSURE

Pressures, sometimes of an enormous magnitude develop during the swelling of an imbibing substance. Such pressures only become evident if the imbibant is confined in some way during the process of imbibition. One common method of demonstrating the development of pressures during imbibition is as follows:

A glass funnel, lined with filter paper is partly filled with moist plaster of Paris paste. The surface of the plaster of Paris is then strewn with a number of pea seeds, after which more paste is added until the funnel is full. In a few minutes the moist matrix containing the seeds will have set. The resulting solid cone is removed from the funnel and its base immersed in a dish of water. Water moves upward through the porous gypsum cone by capillarity and permits a continued imbibition by the pea seeds. Within a few hours the pressure developed by the swelling pea seeds is sufficiently great to rupture the gypsum block.The magnitude of the pressure developed during imbibition may be measured quantitatively in two different ways. It may be counterbalanced by a mechanical pressure, the magnitude of which becomes a measure of the imbibition pressure developed. Such a method was used by Reinke as long ago as 1879. He stacked disks of dried fronds of Laminaria (a sea weed) in a hollow metal cylinder, and inserted above the disks a metal piston bearing a platform at it's upper end. Water was then brought into contact with the dry disks. By placing upon the platform weights of sufficient mass to just prevent swelling of the kelp, the magnitude of the imbibition pressure was determined. A second method of measuring this quantity relies upon counterbalancing the imbibition pressure of a substance by means of a solution of high osmotic pressure. The pressure developed by imbibition may be regarded as due to the diffusion pressure of the entering water. If the imbibing substance be immersed in a solution in which the diffusion pressure deficit (osmotic pressure) is sufficiently great to prevent any movement of water into that substance the osmotic pressure of that solution is a measure of the imbibition pressure which would be developed in that imbibant. A solution with an osmotic pressure, of nearly a thousand atmospheres was required to prevent any imbibition of water by cocklebui seeds. As determined by either of the two methods just described the imbibition pressure of an imbibant is analogous to the osmotic pressure of a solution, i.e. it is a measure of the maximum imbibition pressure which the imbibant can develop.

Chapter 10

Permeability

According to the principles of diffusion, the molecules and ions of all substances in solution, and even colloidal micelles, tend to attain an equal distribution in terms of concentration through all parts of a system. The same is true of gases, and the osmotic movement of water, as we have seen earlier is a response to the same general principle. In living organisms, however, such an equality of distribution of the particles of a solute among the various cells, or between cells and a liquid environment, is seldom attained. There is a marked heterogeneity in the distribution of solutes among the different cells in tissues, in entire organisms, and even among the different parts of the same cell. The diffusion of substances in solution into or out of plant cells is limited and modified in a number of ways. The diversity of the distribution of solutes within the plant body is a result of conditions which modify or check their free diffusion. The permeability of the cell, or some part thereof, is one of the most important factors influencing the rate with which molecules and ions diffuse into or out of plant cells. Plant cells and parts of plant cells such as nuclei and plastids are bounded by membranes which are notable for their differences in permeability to different substances. The permeability of a membrane to a given substance is measured by the rate at which that substance will passthrough that membrane under a given driving force. In the simplest cases, this driving force depends upon the steepness of the concentration or diffusion pressure gradient of the diffusing substance across the membrane.The concept of permeability is inseparable from the idea of a membráne.Permeability, it should be clearly understood, is a property of the membrane and not of the substance which diffuses through it. Common examples of non-living membranes, some of which have already been described in previous chapters, are thin sheets or layers of rubber, collodion, parchment paper, cellophane, gelatin, and copper ferrocyanide. Liquids may also serve as membranes under certain conditions, as an example described later in this chapter shows.

Most biologically important membranes are of the type characterized as differentially permeable. Different substances diffuse through such membranes at different rates. They may be, and often are, impermeable or virtually so to some substances, while others pass through them readily. Such membranes are sometimes termed selectively permeable or semi-permeable, but these two terms are not as truly descriptive of their properties as the designation differentially permeable.

THE MEMBRANES OF PLANT CELLS

With very few exceptions the protoplasm of plant cells is enclosed by a definite, more or less rigid cell wall. This cell wall fulfills all the criteria for being considered a membrane. Through it must pass all substances moving either into or out of a cell. The cell wall membrane is sometimes called the non-living membrane of a cell but the assumption that all cell walls are entirely non-living is scarcely a validone. The plasmodesms, at least, must be considered in any problem involving the permeability of the cell walls of living plants.

Lining the cell wall of all mature plant cells is the layer of cytoplasm. The cytoplasm, or parts thereof, comprises the second membrane through which substances entering or leaving the vacuole of a cell must penetrate. This layer is sometimes called the living membrane of the cell. Both theoretical considerations and experimentally observed facts support the view that the two limiting layers of the cytoplasm possess different physicochemical properties from the interior of the cytoplasm and actually may be regarded as distinct membranes. The cytoplasmic membrane adjacent to the cell wall is called the plasmalemma; that enclosing the vacuole the tonoplast or vacuolar membrane. Since the boundaries between the cytoplasm and the cell wall and the cytoplasm and the vacuole constitute interfaces, protoplasmic ingredients which lower interfacial tension probably become more concentrated in these layers than in the interior of the cytoplasm. For this reason alone the postulation that the cytoplasmic membranes possess different properties from the interior cytoplasm seems justified.

Results from several types of experimental investigations support this hypothesis. The immiscibility of protoplasm with water, already discussed, appears to be at least partly due to the fact that protoplasm is coated with a layer of material which is insoluble in water. By means of micropipettes certain non-toxic dyes can be injected into the body of the cytoplasm through which they will spread rapidly. They do not, however, pass through either the plasmalemma or the tonoplast. Neither will they pass into the interior of the cytoplasm from a solution bathing the cell, nor from the vacuole if the latter is injected with the dye. This is direct evidence that the constitution of the plasmalemma and the tonoplast is different from that of the interior cytoplasm. The presence of cytoplasmic membranes can also be demonstrated by the manipulation of living protoplasm with fine glass needles. It has been found possible to withdraw a plasmolyzed protoplast from its cell wall and to strip off the protoplasm from the vacuole leaving the latter as a free floating sac full of cell sap enclosed by a delicate membrane-the tonoplast.Most evidence indicates that the tonoplast and probably also the plasmalemma are thin liquid films which are immiscible with water. A complete picture of the permeability of a cell to different. Removal of the outer cytoplasm compounds can be drawn only in from the tonoplast and vacuole of a plasmo terms of the permeability of both lyzed cell.

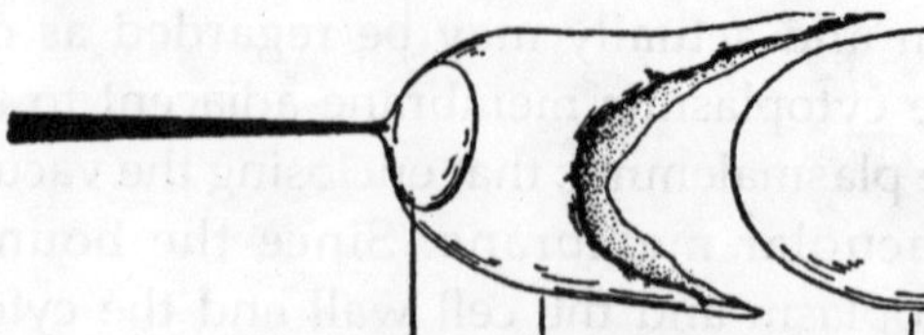

Fig. Removal of the outer cytoplasm from the tonoplast and vacuole of a plasmolyzed cell

In entering a cell a substance must first pass through the various layers of the cell wall, thence in turn through the plasmalemma, the interior cytoplasm, and the tonoplast before reaching the vacuole. Although some investigators prefer to refer all permeability phenomena of plant cells to the cytoplasm as a whole, the evidence that the plasmalemma and the tonoplast exist as distinct membranes with the property of differential permeability must be regarded as very substantial. It is also customary to speak of certain plant structures composed of one or more layers of cells as membranes. Examples are the seed coats of many seeds and fruits (especially grains) epidermal layers such as those peeled from the scales of an onion bulb, the skin of potatoes, etc. Such membranes may be termed for convenience multicellular membranes. They maybe either living, as for example onion epidermis, or non-living, as for example many seed coats.

THE PERMEABILITY OF THE CYTOPLASMIC MEMBRANES

The cytoplasmic membranes are differentially permeable membranes par excellence. The term cytoplasmic membranes will be employed in a loose sense to refer to the entire cytoplasmic system of membranes: plasmalemma plus interior cytoplasm plus tonoplast. Actually many substances entering plant cells are intercepted and utilized in one way or another in the cytoplasm, and never diffuse through the tonoplast. Similarly compounds synthesized in the cytoplasm may pass out of the cell without crossing the tonoplast. Hence the plasmalemma is often regarded as the most important unit in the cytoplasmic system of membranes. A pertinent concept as a background for a discussion of the permeability of the cytoplasmic membranes is the distinction between polar and non polar compounds and groups. Water, electrolytes, carbohydrates, and amino acids are typical strongly polar compounds. This polarity is fundamentally an electrical phenomenon and is due to the grouping of the atoms in a molecule, as well as to the actual structure of its constituent atoms in terms of the spatial arrangement of protons and electrons. A polar molecule can be likened to a tiny magnet since it possesses certain regions which are electro-positive, while other regions are electro-negative. This property of polarity is more strongly marked in some species of molecules than in others. Electrolytes possess the most strongly polarized of all molecules; the production of ions by such compounds may be considered a manifestation of a marked polarity within the molecules. Polar compounds are in general, more soluble in polar solvents, of which water is the most familiar example, thanin non-polar solvents.

The most typical non-polar molecule which is polar at one end; non-polar at the compounds are the hydrocarbons of both the other, chain and ring types, such as methane (CH4) and benzene (C_6H). Non-polar compounds are much more soluble in non-polar solvents such as benzene, chloroform, and ether than in polar solvents such as water.

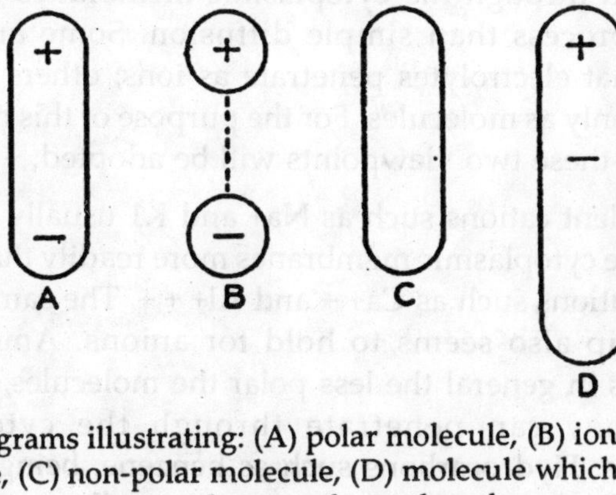

Fig. Diagrams illustrating: (A) polar molecule, (B) ionized polar molecule, (C) non-polar molecule, (D) molecule which is polar at one end; non-polar at the other

The concept of polarity applies not only to molecules as a whole, but to certain atomic groupings within molecules. Such groups as CH_3, C_2H_5, etc., are non-polar, while COOH, OH, NH_2, CHO, CN are the more familiar of the polar groupings. The molecules of many types of compounds, as for example the organic acids, contain both polar and non-polar groups. One end of an acetic acid (CH_3COOH) molecule, for example, is decidedly polar, while the other is essentially non-polar. In general, the greater the number of non-polar groups which a molecule contains in proportion to the number of polar groups present the more predominant will be the non-polar properties of the molecule and vice versa. One of the few important generalizations which may be made regarding the permeability of the cytoplasmic membranes is that they are usually much more permeable to predominantly non-polar than to predominantly polar molecules.

PERMEABILITY OF THE CYTOPLASMIC MEMBRANES

Electrolytes are all strongly polar compounds, hence it is not surprising to find that the cytoplasmic membranes of many plant cells are not readily permeable to them. However, as described later in this chapter electrolytes often enter certain kinds of plant cells readily. The actual mechanics of their penetration through the cytoplasmic membranes is a more complex process than simple diffusion. Some authorities contend that electrolytes penetrate as ions; others that they penetrate only as molecules. For the purpose of this discussion the first of these two viewpoints will be adopted.

Univalent cations such as Na+ and K1 usually penetrate through the cytoplasmic membranes more readily than bi- and trivalent cations such as Ca+ + and Al+ + +. The same general relationship also seems to hold for anions. Among non-electrolytes in general the less polar the molecules, the more readily they can penetrate through the cytoplasmic membranes. Hydrocarbons such as benzene, being the least polar of all compounds, penetrate into cells very rapidly, although in only very small quantities, because of their low solubility in water.

The cytoplasmic membranes are usually quite permeable to the gases carbon dioxide and oxygen when in the dissolved state. This is in accord with the general principle since the molecules of these gases are essentially non polar. When carbon dioxide dissolves in water a large proportion (more than 50 per cent) of the molecules present react with the water, forming carbonic acid (H_2C0_3). This is a complicating factor which must be considered in any evaluation of processes involving the diffusion of dissolved carbon dioxide. Most organic compounds containing only one polar group such as OHCOOH, or NH_2 diffuse across the cytoplasmic membranes less rapidly than those containing no polar groups, but more rapidly than compounds containing two polar groups. Further increase in the number of polar groups present in a molecule still further decreases the penetrability of the molecule. Glycerol ($C_3H_5(OH)$ for example, an alcohol containing three polar OH groups, penetrates relatively slowly, while monohydric alcohols such as methyl (CH_3OH) and ethyl (C_2H_5OH) penetrate through the cytoplasmic membranes very rapidly.

Most studies indicate that the cytoplasmic membranes of most kinds ofcells are usually relatively impermeable to sugars. Impermeability to sucrose is especially marked. Since molecules of sugars contain a large number of polar groups this is in accord with the general conception of the relation of the polarity of molecules to the permeability of the cytoplasmic membranes to them. However, as in the case of electrolytes, there are some kinds of cells which sugars enter relatively rapidly. Amino acids another extremely important group ofcompounds physiologically, also penetrate through the cytoplasmic membranes of most cells relatively slowly. Since each amino acid molecule contains oneor more NH_2 groups and one or more COGH groups the behaviour of these compounds is in accord with the general principle.

Water molecules, although decidedly polar, are an exception to the general rule, usually penetrating through the cytoplasmic membranes with relative ease. This is probably possible only because of the relatively small size of the water molecule. That cytoplasmic membranes are often, at least, quite permeable to water is shown by the frequent rapid loss or gain of water by plant cells.

Another important difference between permeability to polar compounds and to non-polar compounds is that ease of penetration of the former shows a close correlation with particle size, while this is not true of the latter. As among different ions of the same electrostatic charge, for example, the smaller ions (due account being taken of hydration) seem to penetrate more rapidly. No such correlation is evident between particle size and permeability among molecules of a predominantly non-polar structure. In fact, exactly the reverse relation sometimes holds. For example, in the saturated series of fatty acids the larger the number of non-polar groups present, i.e. the greater the size of the molecule, at least up to a certain limit, the greater the permeability of the cytoplasmic membranes to the acid. According to Loeb the order of penetrability of the saturated fatty acids into sea urchin eggs is asfollows, the most rapidly penetrating acid being given first: nonylic ($CH_3(CH_2)_7GOGH$), caprylic ($GH[(GH_2)_6$ $GOGH$), caproic ($GH_3(CH_2)_4$ $GOGH$), butyric (GHy $(CH_2)_2GOGH$), proprionic (CH_3GH_2GOGH), acetic (GH_3 $GOGH$) and formic ($H.$ $GOGH$).

Permeability of the cytoplasmic membranes to polar compounds differs from permeability to non-polar compounds in one other respect. Permeability to the former is markedly influenced by external factors and by internal changes in the membrane. The same membrane apparently may vary all the way from a condition of marked permeability to certain polar solutes to a condition of almost complete impermeability to the same solutes. Permeability to non-polar solutes, on the other hand, is not only relatively high, but much more nearly constant, and much less influenced by environmental factors.

FACTORS AFFECTING THE PERMEABILITY OF THE CYTOPLASMIC MEMBRANES

A number of factors of external origin may influence the permeability of the cytoplasmic membranes. Only the more important of these will be considered. As the preceding discussion has indicated environmental factors influence the permeability of the cytoplasmic membranes to polar compounds more markedly than to non-polar compounds.

TEMPERATURE

The apparent permeability of membranes to many substances increases with temperature. In part this effect may be due to increased kinetic activity of the molecules which pass through the membrane rather than to any effect of temperature on the membrane itself. A given substance will penetrate through a membrane more and more rapidly as the temperature is raised, until the membrane itself is destroyed, or profoundly modified by the temperature attained. In living membranes the upper limit of temperature which results in the destruction of the membrane is the thermal death point of the protoplasm which in most active cells usually lies between 50 and 600 C. At this point a sudden, irreversible increase in permeability occurs. This is due to death of the protoplasm and consequent destruction of the property of differential permeability in the cytoplasmic membranes. This phenomenon may be observed readily by heating a small piece of red beet tissue in a test tube of water. As soon as the thermal death point of the protoplasm is exceeded, the red pigment dissolved in the cell sap (an anthocyanin), to which the living cytoplasmic membranes are impermeable, rapidly diffuses out into the water.

Temperatures low enough to induce ice formation in plant tissues also result in an increase in the permeability of the cells of many (but by no means all) tissues, due to the resultant destruction of the protoplasm. This effect can be conveniently illustrated by freezing a smallblock of red beet tissue and immersing it in water as soon as it has thawed. The almost immediate outward diffusion of anthocyanins which ensues is evidence of the destruction of the cytoplasmic membranes.

ELECTROLYTES

According to Osterhout investigations made on the tissue of Larninaria (a kelp) by the conductivity method 1 indicate that the cations Li^+, Na^+, K^+, NH^{4+}, Cs1, and Rb^+ all increase permeability. This seems to be true regardless of the anion with which the cations are paired. The cations are arranged from left to right in their probable order of effectiveness in increasing permeability. This, it will be noted, is approximately the order of the lyotropic series.

The initial effect of bivalent and trivalent cations such as Ca^{++}, Ba^{++},Sr^{++}, Mg^{+1}, Fe^{+1}, La^{+11}, Fe^{11+} and Al^{11+} as found by this method In this method permeability is determined by measuring the electrical conductivity of cylinders or stacks of disks of plant tissue. The greater the electrical conductivity, the greater the permeability. Since the conductivity of the tissue is due to the movement of ions through it, this method measures permeability to electrolytes only.

This primary effect is usually followed in time by a secondary increasing effect upon permeability. These results were found when these cations were associated with univalent anions. When the anions are of higher valence the initial decreasing effect upon permeability may be diminished or entirely offset.

Similar effects of univalent and bivalent cations on the permeability of the cytoplasm of the epidermal cells of onion bulb scales to water are recorded by de Haan.Raber (1920), using the same method and type of material as Osterhout studied the effects upon permeability of a number of salts composed of the same cation (Na^+) but different anions. All of these salts were found to result in an increase in permeability. When supplied in equimolar concentrations the order of the increasing effect of the anions upon permeability was: citrate] $P0_4$] tartrate] SO_4 and acetate] Cl] NO_3] Br] I]CNS.

The influence of the four polyvalent anions in increasing permeability was much more marked than the univalent anions. Both of these series follow closely the lyotropic order as usually given for anions. The effect of any electrolyte upon permeability can be analyzed only in terms of the anions and cations into which it dissociates. The results of conductivity experiments seem to indicate that the following general principles hold: If the valency of the anion exceeds that of the cation the salt results in an increase in permeability. If the converse is true the initial effect of the salt is a decrease in permeability; this is followed by a secondary increasing effect. When cation and anion are of the same valence the effect will depend upon the relative size of the ions and other factors. When both ions are univalent, however, the usual effect is an increase in permeability.

The interpretation of these effects of electrolytes upon permeability is not entirely clear, but if we assume the cytoplasmic membrane to be composed of negatively charged micelles, an explanation similar to that given in the preceding chapter for the influence of ions upon imbibition would seem to fit the facts closely.

Not all studies of the effects of ions upon permeability have yielded results in accord with those just discussed. For example Brooks found that sodium, potassium, magnesium, and calcium chlorides in concentrations of 0.0125 to 0.05 molar all check the penetration of the dye dahlia into living cells of Nitella. Similarly 115 in found that dilute solutions of the above salts as well as several others all result in a decrease in the permeability of the cytoplasmic membranes of the cells of various species to sugars. The effects of electrolytes upon permeability of the cytoplasmic membranes probably varies not only with their concentration but also with different types ofcells, and with the same cell under different conditions.

The effects of a mixture of two entirely different salts upon permeability should also be considered. This is illustrated in an experiment described by Bayliss. A slice of freshly cut red beet was immersed in a solution of 0.31M sodium chloride. It was observed that the red pigment gradually diffused out into the solution. Since when similar slices were immersed in water no such outward diffusion of red pigment occurred this showed the effect of sodium chloride in increasing permeability. If, however, enough calcium chloride was added to a portion of the sodium chloride solution before the slice of beet was placed in it to bring the concentration of the former salt to 0.17 M no such outward diffusion of red pigment could be detected. Evidently the presence of Ca^{++} ions in some way offset the usual effect of Na^{+} ions upon permeability. Such an effect of one ion upon another is known as antagonism.

NON ELECTROLYTES

Certain non-electrolytes are known to have effects upon permeability. Among these are the anaesthetics. Exposure of the cytoplasmic membranes to extremely low concentrations of ether or chloroform has no apparent effect upon their permeability; exposure to slightly higher concentrations decreases their permeability. With exposure to still higher concentrations this decreasing effect on permeability is only temporary, and is followed by a rapid increase in permeability which may continue until the value characteristic for the death point of that tissue is reached. This secondary effect of anaesthetics is due to the toxic effect of high concentrations upon the protoplasm resulting in its death after a relatively short period of time. The initial decreasing effect of anaesthetics upon permeability is reversible, but the secondary increasing effect is usually non-reversible.

HYDRATION

Changes in the osmotic or imbibitional relations of the water in other parts of the cell may induce changes in the hydration of the cytoplasmic and cell wall membranes. It is inferred that variations in the permeability of the cytoplasmic membranes occur with shifts in their water content. Such changes in permeability are often presumed to be related to changes induced in the viscosity of the cytoplasm due to variations in its degree of hydration. Extreme desiccation of most cells results in a coagulation of the protoplasm, and a consequent complete loss of the property of differential permeability by the cytoplasmic membranes. In some cells, however, particularly those of drought resistant plants, the cytoplasm may regain its normal condition of differential permeability when water again becomes available after a period of practically complete desiccation.

STIMULATION

This ignorance-concealing term is applied to a number of phenomena which result in an increase in permeability, the exact nature of which are so incompletely understood that they cannot be described in any more fundamental terms. For example, it has been found that the permeability of the egg cells of certain species of invertebrate animals increases when fertilization occurs. Likewise injury or wounding of cells or tissues, whether due to mechanical or other causes, also usually results in an increase in permeability.

THE PERMEABILITY OF THE CELL WALL MEMBRANES

The permeability of the cell wall membranes is greatly influenced by their chemical composition and physical organization. Cell walls composed of cellulose and pectic compounds, such as those in most parenchymatous tissues, are usually quite permeable to water, and, if wet, to many dissolved substances. Water and solutes probably pass through such walls largely if not entirely through the hydrophilic intermicellar material. The plasmodesms also undoubtedly facilitate the movement of water and solutes through the walls of many living plant cells.

That the differentially permeable membrane of such plant cells lies in the cytoplasm is visually shown when the protoplasm of any cell which contains anthocyanins dissolved in the cell sap is destroyed by heating, freezing, or toxic agents. As long as such a cell is alive and uninjured the red pigment does not diffuse out of the vacuole, even if the cell is immersed in water. Immediately upon destruction of the protoplasm, however, outward diffusion of anthocyanins and other solutes occurs through the intact cell wall, indicating that the latter does not appreciably impede their egress from the cells.Lignified cell walls are also quite freely permeable to water and, when wet to solutes as well. Lignification apparently results in little if any decrease in the permeability of the intermicellar material.

Cell walls in which cutin or suberin are present in appreciable quantities on the other hand, are relatively impermeable both to water and solutes. While it appears to be generally true that cell wall membranes which are readily permeable to water are also readily permeable to solutes dissolved in water, a number of differentially permeable cell wall membranes have been discovered and it is probable that differential permeability is more often a property of the cell walls than is generally realised. The coats of a number of different kinds of seeds and fruits have been found to act as differentially permeable membranes. These coats may be several cell layers in thickness but the cells of which they are composed are often dead. Even in those in which the cells are living destruction of the protoplasm by one treatment or another does not destroy the property of differential permeability. Clearly therefore, this property is inherent in the cell walls. Many such membranes are relatively permeable to water but show different degrees of permeability to solutes, ranging all the way from practical impermeability to some to a marked permeability to others. Among the species whose seed or fruit coats have been found to show this property are many of the grains, peach, apple, bean, horse-chestnut, sugar-beet, sunflower and cocklebur. For example, Shull found that the seed coat of the cocklebur was impermeable to ether acetone, chloroform, ethyl alcohol, glycerol, sugars, NaCl, CuSO4, HCl, and tartaric acid among others, but was permeable to a greater or lesser degree to KCl, KOH, NaOH, H_2S0_4, HNO_3, $AgNO_3$, $NaNO_3$, acetic, citric, and lactic acids, as well as some others.

RELATION BETWEEN PERMEABILITY AND THE RATE OF ABSORPTION

There is not necessarily any correlation between the permeability of the cell membranes to a given compound and the rate at which that substance will enter the cell from other cells or a liquid medium. The permeability of the cell membranes does not control the rate at which compounds pass either into or out of a cell but is merely one factor which influences their rate of movement through the cell wall and the cytoplasm. For example, the membranes of a cell may be readily permeable to a given substance which nevertheless may enter very slowly because the vacuolar concentration of that substance is already relatively high. Another substance, to which the membranes are relatively impermeable, may move into the cell continuously because it is being utilized in some metabolic process within the cell.

Recent work on the absorption of ions by plant cells indicates that the results of many studies upon permeability to electrolytes are somewhat misleading. In actively metabolizing cells ions often accumulate until their vacuolar concentration is much greater than in the outside medium. The penetration of electrolytes into such cells often occurs at a relatively rapid rate, in spite of much evidence which indicates that the cytoplasmic membranes are not very permeable to them. Similarly there is evidence that sugars sometimes move into at least certain kinds of cells at a relatively rapid rate. A probable explanation of this apparent discrepancy is that most studies of permeability have been made upon cells which do not possess the capacity of absorbing electrolytes or other polar compounds at a rapid rate. This capacity seems to be restricted largely, if not entirely, to meristematic or otherwise rapidly metabolizing plant cells.

THEORIES OF MEMBRANE PERMEABILITY

A number of hypotheses have been proposed as an explanation of the mechanism of differential permeability in both inorganic and living membranes. The discussion will be limited largely to the bearing of these theories upon the permeability of the cytoplasmic membranes. The two theories which have been most favourably regarded in all discussions of this topic are the sieve theory and the solubility theory. A number of variants of each of these theories has been proposed.

SIEVE THEORY

Probably the simplest picture which can be conceived of the operation of a membrane is that it works as a molecular sieve. The pores in the membrane may be imagined to be of such dimensions that micelles molecules or ions below a certain size can diffuse through them while those exceeding this size limit cannot pass through. Certain types of inorganic membranes such as collodion, parchment paper, and copper ferrocyanide undoubtedly owe their property of differential permeability largely to such a sieve-like action.

Although a sieve theory of the permeability of the cytoplasmic memb rane sis strongly advocated by some authorities, notably Bayliss there are some very grave objections to the acceptance of any such hypothesis as a complete explanation of this phenomenon. It is difficult to conceive of a sieve like structure of the cytoplasmic membranes, except insofar as this is imparted by the normal intermolecular distances which are of course, very minute. A more serious difficulty, however, is found in the fact that the permeability of the cytoplasmic membranes to certain classes of compounds actually increases with increase in the size of the molecule, at least up to a certain point. This is true in certain homologous series of organic compounds already cited.

SOLUBILITY THEORIES

A second generally advocated theory is that membranes are permeable to substances which will dissolve in them and impermeable to others. The operation of an inorganic membrane of this type can be readily demonstrated by a simple experiment. A test tube is about half filled with chloroform on top of which is introduced a very thin layer of water, and then nearly filled with ether. The water may be coloured faintly with a water soluble dye such as methylene blue in order to aid in distinguishing it from the other liquids. After stoppering, the tube is clamped in an upright position and the position of the water meniscus is marked accurately. A second tube is prepared in exactly the same way, except that xylene is substituted for ether as the upper layer of liquid. If examined after a period of several days, it will be found that in the first test tube the level of the water meniscus has risen, while in the second it has fallen, although the distance through which the meniscus will move will not be as great in the second tube as in the first. In the first tube ether is diffusing through the water membrane more rapidly than chloroform, hence the volume of liquid below the membrane is constantly increasing, and the layer of water rises. In the second test tube the chloroform diffuses through the water membrane more rapidly than the xylene hence the position of the layer of water is lowered. This behaviour is in accordance with the known solubilities of these three compounds in water; ether is the most soluble, chloroform next, and xylene the least. Evidently in this experiment permeability of the water membrane to these three compounds is directly correlated with their solubility in water.

The most familiar form of the solubility theory as applied to cytoplasmic membranes is the lipoid solubility theory first proposed by Overton.In brief this theory holds that substances soluble in lipoids pass readily through the cytoplasmic membranes while others do not..The term lipoid is used in a loose way to refer to fats and fat-like compounds. In recent usage the term lipid has been rather generally substituted for lipoid. Several lines of evidence appear to favour this theory. Among these are:

1. The non-miscibility of protoplasm with water, which probably indicates the presence of a film of fatty material on the outer surface of the protoplasm,
2. The fact that most lipoids lower surface tension, and hence theoretically would accumulate at interfaces, and

3. The high resistance of protoplasm to the passage of an electrical current, which indicates that it contains layers through which ions do not diffuse readily. A fourth line of evidence for this theory comes from Overtons own experiments. He foundan almost perfect correlation between the lipoid solubility of hundreds oforganic compounds and the permeability of the cytoplasmic membranes to them. On the other hand, sugars, salts, and other lipoid-insoluble substances showed practically no penetration into the cells at all. We now know that the terms lipoid-soluble and lipoid-insoluble correspond very closely to the modern terms of non-polar and polar, respectively.

While Overtons experiments seemed to show that the cytoplasmic membranes are relatively impermeable to polar compounds, later work has not all supported this generalization. It is true that polar compounds often penetrate with much greater difficulty than non-polar compounds and that the permeability of a given membrane to them may fluctuate, but the membranes are at times at least, permeable to them. Moreover on a priori grounds it is evident that the membranes must be permeable to sugars, salts, etc. at least part of the time, since in the majority of cells most of these compounds present must have entered the cell at some time or another during its life history. Some investigators hold that the lipoid solubility theory is essentially a correct representation of the mechanism of cytoplasmic permeability. Polar compounds would not be completely insoluble even in a lipoidal layer, and hence could penetrate through the membrane at relatively slow rates. Osterhout on siders that electrolytes must diffuse through this layer principally in the form of molecules as their dissociation when dissolved in a fat like liquid would be very slight. Other authorities are not convinced that this theory presents a complete picture of the basis for differential permeability in the cytoplasm. Observed rates of penetration of some polar compounds seem too great to be accounted for satisfactorily by the lipoid-solubility hypothesis. Modifications of this theory have therefore been suggested. For instance Nathansohn has proposed a theory which postulates a membrane composed of a mosaic of proteins and lipoids. He assumed that polar (lipoid-insoluble) compounds could diffuse through the more or less hydrated protein portion of the membrane. Permeability to the latter class of compounds might vary with changes in the hydration of the proteins. Although this view is ahriost purely speculative some properties of the membrane, such as its elasticity, suggest the presence of proteins as one of its constituents.

ADSORPTION AND CHEMICAL REACTION THEORIES

It has been suggested that the permeability to non-polar compounds is correlated more closely withthe adsorptive capacity of the particles of the membrane for them than to their solubility in membrane constituents. In brief this adsorption theory would hold that the more strongly molecules are adsorbed, the more readily they will pass through the membrane. The possible mechanics of the transport of adsorbed molecules across a membrane is described later.

Chemical reactions may be involved in the transport of certain types of solutes across the cytoplasmic membranes. A penetrating molecule or ion according to such a chemical reaction theory, is supposed to combine with some substance at the outer boundary of the membrane, thus forming an intermediate compound, in the form of which it is presumed to move across the membrane. Once the other boundary of the membrane has been reached a second chemical reaction is supposed to occur resulting in release of the ion or molecule on the interior of the membrane.

Whether the penetration of a molecule depends upon solution, adsorption or chemical reaction the probability is strong that the cytoplasm itself actively participates in some manner in the movement of substances through it. For example, after molecules of a penetrating substance dissolve in, are adsorbed by, or combine chemically with micelles of the membrane, their penetration may be facilitated by kinetic activity of the membrane particles. A micelle might act in the capacity of a ferry boat, transporting a cargo of one or more molecules from one side of the membrane to the other. Once across the membrane many of the passenger molecules would escape because of their relatively low concentration in the medium in contact with that side of the membrane. A seemingly more probable hypothesis is that the cargoes of adsorbed, dissolved, or chemically bound molecules are not carried all the way across the membrane by one micelle, but pass from particle to particle as impacts occur between them. Each micelle would, according to this hypothesis carry a given molecule only for a very short portion of its journey across themembrane.

THE MECHANISM OF CHANGES IN MEMBRANE PERMEABILITY

The permeability of the cytoplasmic membranes fluctuates continuously as a normal feature of cellular activity. Some changes in the permeability of these membranes are induced by variations in factors of external origin as previously discussed, but others are due to unrecognized factors in the internal environment of the membranes. Furthermore, the living membrane of a given cell may at times be more permeable to some substances in certain regions thanin others, whereas a short time later the distribution of the relatively permeable are as may be very different.

Various suggestions regarding the mechanisms by which changes in the permeability of the cytoplasmic membranes can take place have been proposed. Among these are:

1. Changes in the viscosity of the membrane,
2. Reversal of phase (on the assumption that the membrane has an emulsion-like structure), and
3. Changes in the proportion of water in different phases of the colloidal system which is supposed to constitute the membrane. Since all of these hypotheses are highly speculative any detailed consideration of them in an introductory discussion would be unjustified.

Chapter 11

The Osmotic Quantities of Plant Cells

The Plant Cell as an Osmotic System. We have already become acquainted with the typical plant cell as a three dimensional structure, limited by cell walls; these walls in turn being lined with a layer of protoplasm which encloses a vacuole filled with cell sap. The protoplasm is normally held against the cell wall by a pressure exerted by the cell sap. The cell walls of most living cells are elastic within limits, but possess sufficient tensile strength so they do not commonly rupture. The cytoplasm, although usually possessing a high degree of fluidity, is usually much less elastic than the walls.The cell wall membranes of most living plant cells appear to be quite freely permeable to water and to most substances dissolved in it. The cytoplasmic membranes, on the contrary, almost invariably possess the property of differential permeability. Water and certain solutes pass through them much more readily than certain other solutes.

PLASMOLYSIS

If a plant cell which is at least partially distended is immersed in a hypertonic solution a characteristic series of changes takes place in the appearance of that cell. The first detectable occurrence is a gradual shrinkage in the volume of the entire cell due to outward osmosis and a consequent reduction in the pressure exerted by the cell sap against the protoplasm and cell wall. This shrinkage in volume can be detected in many cells by actual measurement under a microscope, although there are some types of plant cells in which little or no change in volume occurs under such conditions. There is a lower limit to the elasticity of the cell wall however, and when this is attained no further contraction in the volume of the cell will occur. Since the cell wall is quite freely permeable to the water and the solutes of the surrounding solution, and hence the hypertonic solutionis in contact with the outer surface of the protoplast, the cell sap will continue to lose water by outward osmosis, just as if no cell wall were present.

With reference to a. given solution, in this case the cell sap, a hypertonic solution is one with a higher osmotic pressure than the reference solution. Similarly an isotonic solution is one with an osmotic pressure equal to that of the reference solution, while a hypotonic solution is one of lesser osmotic pressure. Cell wall has ceased as, unlike the latter, the protoplasm has no lower limit of elasticity. At this point, therefore, the protoplasmic layer will begin to separate from the cell wall. If the hypertonic solution is strong enough this separation of the protoplasm from the cell wall will become very pronounced.

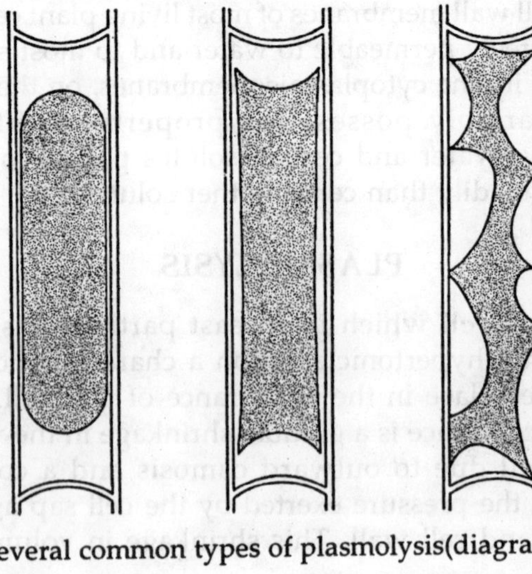

Fig. Several common types of plasmolysis(diagrammatic)

In some types of cells the protoplasm will ball up into a more or less spherical mass within the cell. This phenomenon is called plasmolysis. The pattern followed by the shrinkage of the protoplasm in plasmolysis is more or less typical for each kind of cell, although it may be modified somewhat depending upon physico-chemical conditions within the protoplasm kind of solute used in the plasmolyzing solution, etc The space between the cell wall and the protoplasm will be filled, after the separation of the latter from the wall, with the external solution.

If a plasmolyzed cell is immersed in water it will slowly recover and regain a turgid state, due to osmotic movement of water into the cell sap.Similarly if immersed in a solution hypotonic to the cell sap recovery will also ensue, but the degree of turgidity attained will be less than if the cell is immersed in pure water. Rapid deplasmolys is, however, results in the death of many kinds of plant cells.

In the preceding discussion plant cells have been considered purely a sosmotic systems and we will continue to so regard them throughout the chapter. Although this has been the prevailing concept at least regarding mature, thin walled plant cells with prominent vacuoles the possibility that other mechanisms may also operate in the cell-to-cell movement of water should not be disregarded. The cytoplasm may participate actively in the passage of water into or out of plant cells. Bennett-Clark, et al. present evidence which they believe indicates that cytoplasm may secrete water into the vacuoles of some types of cells. Electro-osmotic phenomena may also be involved in the cell-to-cell movement of water. Furthermore, in some types of cells the vacuoles are very small or are entirely filled with mucilaginous materials of one type or another. In such cells imbibitional phenomena probably play a greater proportionate role in the movement of water than in cells in which the vacuole occupies most of the cell volume.

METHODS OF DETERMINING THE OSMOTIC PRESSURES OF PLANT CELLS AND TISSUES

The plasmolytic method and the cryoscopic method are in common use for determining the osmotic pressure of plant cells and tissues. Both methods have many limitations and neither is as accurate as might be desired. DeVries was the first to use the plasmolytic method which in principle is very simple. A series of solutions (usually of sucrose) graded according to volume molar concentration is first prepared. The range of concentrations to be used depends upon the tissue to be studied. Comparable strips or sections of the tissue are then immersed in each solution and left until an osmotic equilibrium is attained. After immersion in the solution the pieces of tissue are observed under a microscope. In the stronger solutions it will be found that all of the cells are severely plasmolyzed while in the weaker ones little or no sign of plasmolysis can be detected. Somewhere in the series will be found a solution in which about one-half of the cells are not plasmolyzed, and about one-half are more or less completely plasmolyzed. The average osmotic pressure in the cells of the tissue under investigation is considered to be equal to the osmotic pressure of the solution in which this condition obtains.

The value obtained by the procedure just outlined is called the osmotic pressure at incipient plasmolysis. This value is often greater than the true osmotic pressure of the cells since plasmolysis of many kinds of cells is preceded by a shrinkage in their total volume. A cell which has an osmotic pressure of 15 atmos. at incipient plasmolysis might have had an osmotic pressure of (for example) vz or 13 atmos. in its normally distended state. Certain more precise methods of determining the osmotic pressure of cells by the plasmolytic method, which take into consideration their volume changes have also been devised. If incipient plasmolysis is determined to occur for cells of a certain tissue in a sucrose solution with a volume molar concentration of M then the volume molar concentration of sucrose equivalent to that of the cell sap at its original volume

M (i.e. before the shrinkage due toplasmolysis),is related to M by the following equation: M1V1Mv in which V and V1 represent the volume of the cell in its original condition,and at incipient plasmolysis respectively. Having determined M1 the corresponding molarity for the cell sap of the normal cell can be readily calculated.

The practical difficulty of all such methods is in measuring accurately the volume changes which occur in cells during plasmolysis. The usual practice is to measure the change in the length (sometimes also in the breadth) and to assume that such measurements are proportional to the change in volume of the cell. For some types of cells this assumption is probably valid but with others it may result in serious errors. Formerly solutions of a number of different compounds, including certain electrolytes, were used in plasmolytic determinations, but at the present time sucrose solutions are almost universally employed. They are preferred because:

1. Sucrose is apparently non-toxic to the protoplasm,
2. It does not penetrate into plant cells at a very appreciable rate
3. It apparently has little or no influence on membrane permeability, and
4. The exact osmotic pressures of different molar concentrations of sucrose are known with greater accuracy than for most other solutions.

Even when sucrose is used as a plasmolyzing agent the method is subject to a number of errors, of which the following are the most serious:

1. An outward diffusion of solutes into the plasmolyzing solution may occur during the determination thus resulting in an apparent osmotic pressure which is less than the true one.
2. In many plant cells it is difficult to observe the initial stages in the separation of the protoplasm from the cell wall during plasmolysis. This is most easily observed in cells with a coloured sap or in which

the cytoplasm contains chloroplasts, hence they are generally used to demonstrate plasmolysis.

3. The cytoplasm sometimes adheres so strongly to the cell wall that a solution of considerably greater concentration than one just barely hypertonic to the cell sap may be required to bring about plasmolysis. As a result of this source of error the apparent osmotic pressure is greater than the true one.
4. If the osmotic pressure of the cells of organs of the higher plants is to be measured by this method, strips of tissue thin enough to permit its examination under a microscope must be cut out of the organ. The resultant mechanical shock to the cells may have profound influences on conditions within them, which may in turn affect their osmotic pressures.
5. The time required for an equilibrium to be obtained between a cell and a solution which just causes incipient plasmolysis is often so great that various types of changes may take place in the cell during this period and render the determination of dubious accuracy.

When other plasmolyzing agents than sucrose are used other sources of error are often introduced into the determination in addition to those listed above. It is doubtful if measurements of osmotic pressure by the plasmolytic method are ever very accurate except perhaps for types of cells which are inherently well adapted to the method. The principal advantage of this method is, that with rare exceptions, it is the only feasible one available for measuring directly the osmotic pressures of living plant cells.The principle of the cryoscopic method for determining osmotic pressures has already been described in earlier chapter. Briefly, as usually applied to plants, the method consists in expressing the juice from a sample of plant tissue and determining its freezing point depression with a suitable thermometer or a thermocouple system. The osmotic pressure can then be calculated from the freezing point depression by the usual equation. It has been found to be impossible to obtain homogeneous samples of sap from plant tissues unless they are first treated in such a way as to kill the cells. Consequently the tissue is usually first subjected to freezing, heating, grinding or exposure to toxic vapours before the sap is expressed.

Doubtless the expression under high pressure of the sap from tissues which have previously been subjected to such treatments has a modifying effect upon the properties of the expressed sap of most tissues as compared with the original cell sap. Furthermore the sap obtained represents a mixture of contributions from all of the cells living and dead in the tissues. At the best, determinations of osmotic pressure made on such saps can represent no more than an average of the osmotic pressures of all the cells in the tissue. In spite of the possibility of a considerable modification in the properties of the sap inherent in this method, there is probably a consistent correlation between values obtained in this way and the mean osmotic pressure of the cell sap of the cells in a tissue.

Magnitude of the Osmotic Pressures in Plant Cells. In many discussions of the osmotic pressure of plant cells no distinction is made between the osmotic pressure of the cells at incipient plasmolysis and their osmotic pressure under more or less turgid conditions. When the term osmotic pressure is employed without qualification it can be assumed to refer to the osmotic pressure of cells in their normally distended condition. It is generally considered that the results of determinations by the cryoscopic method represent approximately the osmotic pressure of the cells in this condition.

Different organs or tissues of the same plant may exhibit a wide range of osmotic pressures. Even similar organs on the same plant leaves for example may vary considerably among themselves in the average osmotic pressure of their cells. Furthermore the tissues within the same organ usually show a considerable variation in osmotic pressures. The mesophyll cells of most leaves, for example, show higher values than the epidermal cells. In general the osmotic pressures of the plant cells and tissues of the mesic species of North America vary in their extreme range from a fraction of an atmosphere to about 50 atmos. Most of the values lie, of course, within a narrower part of this range, probably within the limits of a 4 to 30 atmos.span of values. In general the cells of ligneous species of plants as a group appear to have somewhat higher osmotic pressures than herbaceous plants as a group. This generalization cannot of course be extended to a comparison of any individual ligneous species with an individual herbaceous species. The range of osmotic pressures in the leaf tissues of plants which is self-explanatory. It should be clearly understood that values such as those given in this table by no means represent constants. Considerable variation may be found in anyone species depending upon environmental conditions, position of leaf on the plant, etc.

FACTORS DETERMINING THE OSMOTIC PRESSURES OF PLANT CELLS

Any factor which influences either the water content of a plant cell or the solute content of its cell sap will have an effect on the magnitude of the osmotic pressure of that plant cell. The water content of the plant as a whole, and hence of its constituent cells, is controlled principally by the relative rates of transpiration and the absorption of water. The latter process is influenced very markedly by the water content and other conditions prevailing in the soil. Individuals of the same species invariably have a higher osmotic pressure when growing under drought conditions than when provided with a favourable water supply. This is at least partially due to the low water content of the leaves which results when the available soil water supply becomes low. Other factors which are also probably involved are a decrease in the growth rate of the plant which often permits an accumulation of mineral salts and soluble foods, and a shift of the starch 1 soluble carbohydrates It should perhaps be emphasized that the term osmotic pressure of a tissue can possess meaning only in the sense of the average osmotic pressure of the cells composing that tissue.

The solute Content of the Cell sap is Controlled by the specific metabolic processes of the plant and by the absorption of mineral salts by the plant from its environment. The rate of photosynthesis is an important factor indetermining the osmotic pressure of plant cells, particularly those of leaf tissues. The influence of the inherent metabolic processes of any species upon the kinds and concentrations of various types of soluble organic compounds produced, such as simple carbohydrates, organic acids, amino acids, etc. has an exceedingly important effect on the magnitude of the osmotic pressure in any species. Metabolic conditions and their effects upon osmotic pressures may also be altered by environmental conditions. A well-known example of this is the difference in the osmotic pressures of sun and shade leaves on the same plant. The former almost invariably have the higher osmotic pressure due presumably to their greater photosynthetic activity.

The mineral salts which contribute to the osmotic pressures of plant cells are all absorbed from the soil solution, or in aquatic species from the water in which part or all of the plant is immersed. Different species of plants vary greatly in their toleration of high concentrations of mineral salts in the soil. All species can become adjusted within limits to a change in the mineral salt content of the substratum. This adjustment takes the form of an increase in the osmotic pressure of the plant with an increase in the osmotic pressure of the medium from which it obtains its mineral salts.

Other tissues of plants also show an increase in osmotic pressure with increase in the osmotic pressure of the soil solution, but such increases are generally most pronounced in the roots.Species indigenous to alkali soil regions usually have relatively high osmotic pressures. Alkali soils are rich in soluble salts, and the high osmotic pressures found in species native to such soils are due to the absorption of relatively large quantities of mineral salts. In fact the highest recorded osmotic pressure for any species of plant atmos was found in an alkali soil species, A triplex con fertifolia. Salt marsh plants also generally have relatively high osmotic pressures. This is also a correlation with the relatively high salt content of the substratum. Diurnal Variations in the Osmotic Pressures of Plant Tissues. The osmotic pressures of plant tissues fluctuate diurnally. Such variations are more marked in the aerial than in underground portions of plants. Their magnitude and direction are very strongly influenced by the environmental conditions to which the plant is subjected. Daily variations in osmotic pressure have been studied more extensively in the leaves than in any other organ of the plant. The data for the leaves of the prairie grass, Andropogon scoparius were obtained under extreme drought conditions, but similar, although less marked diurnal variations in the osmotic pressure of leaves are undoubtedly of regular occurrence in most species, at least on clear days.

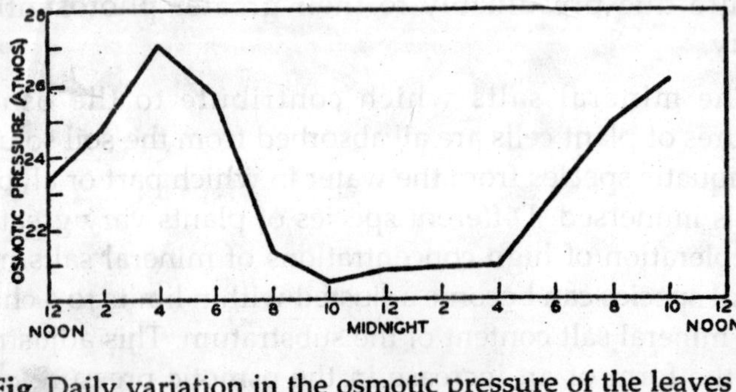

Fig. Daily variation in the osmotic pressure of the leaves of Andropogon scoparius

SEASONAL VARIATIONS IN OSMOTIC PRESSURES

The rather consistent increase in the osmotic pressure of leaves which usually occurs during the daylight hours is probably conditioned primarily by two factors: the accumulation of soluble carbohydrates or other organic compounds resulting directly or indirectly from photosynthesis, and the decrease in the water content of the cells due to an excess rate of transpiration over the rate of absorption of water resulting in a shrinkage in the volume of the cells.The minimum leaf osmotic pressure, which is usually attained between midnight and dawn, probably corresponds to the period when the leaf cells are at their maximum water content, and at their minimum organic solute content due to the continuance of translocation out of the leaves during the night. Seasonal Variations in the Osmotic Pressures of Plant Tissues.

In temperate regions the osmotic pressures of at least the exposed organs of plants are usually higher in the winter than in the summer. Seasonal variations in the osmotic pressures of temperate zone plants have been studied principally in the leaves of evergreen species. Such variations are also known to occur in the exposed woody stems of plants, although information about them is much less complete. The increased winter osmotic pressures are undoubtedly correlated principally with an increase in the soluble carbohydrates in the cell sap due to a shift in the starch soluble carbohydrate equilibrium caused by low temperatures and to a lesser extent to the reduced water content of the leaf tissues which is usually characteristic of the winter condition. In some species the first of these factors is predominant, in other species the second. Similar seasonal variations in osmotic pressure occur in the leaves of evergreen herbaceous species, as for example winter wheat. In many woody species of semi-desert regions the osmotic pressure of the leaves increases markedly during the hot, dry season (May-July) but shows a rapid decrease with the advent of summer rains. The increasing osmotic pressure of the leaves of such species during the dry months is closely correlated with the increasing unavailability of soil water.

THE OSMOTIC QUANTITIES OF PLANT CELLS

Let us suppose three similar cells, each in a state of incipient plasmolysis, and each with a cell sap osmotic pressure of 10 atmos., to be immersed, the first in pure water, the second in a solution with an osmotic pressure of 6 atmos.

and the third in a solution with an osmotic pressure of 10 atmos. The volume of the liquid in which each cell is immersed is supposed to be very large in proportion to the volume of the cell.

In order to simplify the discussion certain assumptions will be made regarding these cells and, unless stated to the contrary, regarding all other cells discussed in the remainder of this chapter. These are:

1. That the cells are at the same temperature,
2. That the cytoplasmic membranes are permeable only to water while the cell walls are freely permeable to both water and solutes and
3. That the walls of all the cells are equally elastic. Furthermore, changes in the osmotic pressure of the cell sap of these three cells due to the changes in the volume of the cells will be disregarded. Diffusion of water into cell A begins immediately upon its immersion.

This inward movement of water is osmosis. The diffusion pressure deficit of the water in the cell sap is JO atmos. (since its osmotic pressure is 10 atmos and it is not subjected to pressure) ; that of the surrounding water zero, hence the diffusion pressure of the external pure water is 10 atmos. greater than that of the water in the cell sap. The entrance of water into the cell exerts a gradually increasing pressure against the protoplasm which is in turn transmitted to the cell wall. This pressure is called the turgor pressure and is the cause of the gradual distension of the cell.

As soon as any turgor pressure is exerted against the walls of the cell they exert a counter pressure the wall pressure against the protoplasm and cell sap. The wall pressure is always equal to but acts in the opposite direction to the turgor pressure. The maximum turgor pressure which can be developed by this particular cell is 10 atmos., since the osmotic pressure of a solution is a measure of the maximum turgor pressure which it can develop. At this point the wall pressure of the cell will also be equal to 10 atmos. The wall pressure of a cell will have exactly the same effect upon the diffusion pressure of the water in the cell sap that a pressure of equal magnitude exerted by a piston would have on water subjected to it. In this cell, therefore, the wall pressure will increase the diffusion pressure of the enclosed cell sap by 10 atmos. Since the initial diffusion pressure deficit of the cell sap was Jo atmos. and the development of a wall pressure of 10 atmos. has raised the diffusion pressure of the water in the cell sap by JO atmos. the diffusion pressure deficit of the water in the cell sap is reduced to zero. Since this is also the diffusion pressure deficit of the surrounding water, a dynamic equilibrium has been established due to the influence of the wall pressure upon the diffusion pressure of the water in the cell sap. When this condition of dynamic equilibrium has been attained, equal numbers of water molecules will be passing across the membrane in both directions per unit of time.

It is the turgor pressure of the cell acting in opposition to its wall pressure which imparts to plant cells their usual rigid, distended condition. This condition is termed turgor, turgidity, or turgescence; the first of these terms being in most general use. Plant cells, although generally turgid, seldom attain a turgor pressure equal to their osmotic pressure. Cells low or entirely lacking in turgor pressure are usually referred to as flaccid. Cell C will remain, according to the conditions prescribed for this experiment, in a state of incipient plasmolysis. The diffusion pressure deficit of the water in the cell sap and in the solution are equal (i.e. their osmotic pressures are equal and neither is under any pressure, except, of course, atmospheric pressure) hence a dynamic equilibrium will be established as soon as the cell is immersed in the solution. Since there is no net movement of water into the cell no turgor pressure and hence no wall pressure will develop, a condition which will obtain in all plant cells in the condition of incipient plasmolysis.

The diffusion pressure deficit of the water in the cell sap of cell B is 10 atmos., that of the water in the surrounding solution 6 atmos. Hence the diffusion A the water in the solution is 4 atmos. greater than that of the water in the cell sap, and there will be a net inward movement of water into the cell. A turgor pressure thus develops within the cell, but at its maximum under these conditions it will attain a value of only 4 atmos. Counterbalancing this turgor pressure will be a wall pressure of 4 atmos. Since the initial diffusion pressure deficit of the water in the cell sap was 10 atmos., imposition of a wall pressure of 4 atmos. reduces this to 6 atmos which is the diffusion pressure deficit (osmotic pressure) of the water in the surrounding solution. Hence the attainment of a turgor pressure of only 4 atmos. results under these conditions in an osmotic equilibrium. Since the magnitude of the turgor pressure developed in cell B will be less than in cell A, the latter cell will be more completely distended than the former. In all three of these cells a dynamic equilibrium has been attained by an adjustment of the diffusion pressure deficit of the water in the cell sap until it is equal to that of the water in the surrounding liquid. This adjustment was attained in each of these cells by a shift in the magnitude of the wall pressure, which is one of the two components determining the diffusion pressure deficit of the water in the cell sap, the other being the osmotic pressure. Unlike a solution exposed to the atmosphere, the diffusion pressure deficit of a cell is not equal to its osmotic pressure except when the cell is flaccid, i.e., possesses a zero wall pressure. The term diffusion pressure deficit of a cell should be considered as an abbreviation for diffusion pressure deficit of the water in the cell sap.

In the interests of simplicity the effects of such volume changes upon the osmotic pressure, etc. of the cell have been disregarded in the discussion up to this point. It is assumed that the wall pressure of the cell increases proportionately with an increase in the volume of the cell. When the cell is completely flaccid its diffusion pressure deficit is equal to its osmotic pressure (2o atmos.) while its turgor pressure and hence its wall pressure are both zero. As the volume of the cell increases, due to an influx of water, the osmotic pressure decreases due to dilution of the cell sap. The turgor pressure and wall pressure increase equally and progressively with increase in the volume of the cell, the diffusion pressure deficit of the cell shows a corresponding but more rapid, progressive decrease. When the cell attains a condition of maximum turgidity (volume 1.5) its turgor pressure and wall pressure are equal to its osmotic pressure while its diffusion pressure deficit has fallen to zero.

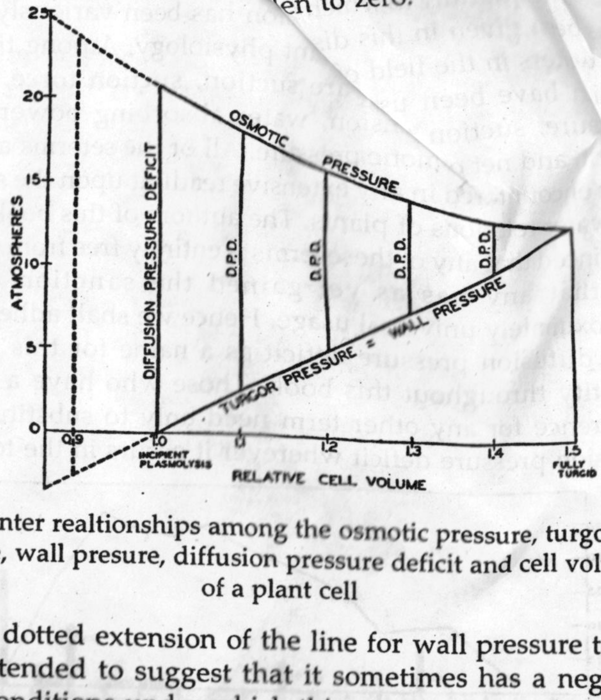

Fig. Inter realtionships among the osmotic pressure, turgor pressure, wall presure, diffusion pressure deficit and cell volume of a plant cell

The dotted extension of the line for wall pressure to the left is intended to suggest that it sometimes has a negative value. Conditions under which this may occur are discussed shortly. The diffusion pressure deficit of a cell in which the wall pressure is negative is greater than its osmotic pressure. In many types of cells volume changes are much less marked and hence they can often be disregarded in generalized considerations of the water relations of plant cells. In some species many of the cells have walls which are totally lacking in elasticity. This is especially true of cells in the tissues of xerophytes and some water plants.

The fundamental relations between the osmotic pressure, wall pressure, turgor pressure, and diffusion pressure deficit

of a pl cell are expressed in the fo simple equations: olution of 6 atmos.
after taining an equilibrium ure of 10 atmos., and a
osmotic ressure, had an osmoti diffusion pressure deficit
wall pr sure of 4 atmos. Henre was 4 atmos.
was 6 tmos and its turgor p

The quantity to which name diffusion pressure deficit has been given in this d ssion has been variously termed by writers in the field plant physiology. Among the terms which have been u d are suction, suction force, suction pressure, suction tension, water absorbing power, turgor deficit, and net osmotic pressure. All of the seterms are likely to be encountered in any extensive reading upon the subjectof the water relations of plants. The authors of this book are not convinced that any of these terms is entirely free from criticism nor that any has as yet gained the sanction of even approximately universal usage. Hence we shall adhere to the term diffusion pressure deficit as a name for this physical quantity throughout this book. Those who have a decided preference for any other term need only to substitute it for diffusion pressure deficit wherever it occurs in the text.

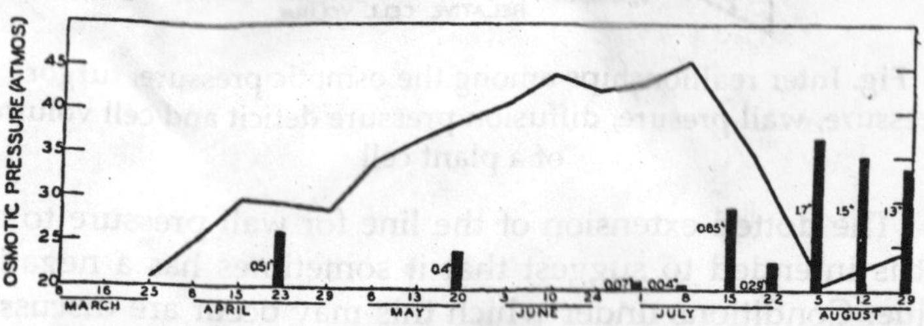

Fig. Seasonao variation in the osmotic pressure of the leaves of the creosote bush under desert conditions

The osmotic pressure, diffusion pressure deficit, turgor pressure, and wall pressure are collectively called the osmotic quantities of plant cells. As previous discussion has shown a full evaluation of these quantities also requires a consideration of the volume changes of plant cells. In plant tissues many of the cells are under a pressure imposed upon them by the surrounding cells. In addition to its own wall pressure the water in such a cell is also subjected to this pressure of external re pressure is just as much a factor in determining deficit ure deficit of a cell as its own wall pressure. essure deficit of such a cell is therefore l considered as an individual unit by d pressure. Hence the equation ermining the diffusion pressure ditions becomes.

Fig. Similar cells immersed i solutions … ent osmotic presures

DIFFUSION PRESSURE DEFICIT OSMOTIC PRESSURE WALL PRESSURE PRESSURE EXERTED BY SURROUNDING CELLS

The water in the vessels and cells of plants frequently passes into a state of tension. Imposition of pressure from an external sourc: raises the diffusion pressure of water. Imposition of a tension (which is equivalent to a negative pressure) has pecisely the opposite effect. In cells and vessels this can only happen if the enclosed water contracts in volume to such a point that theencompassing walls are pulled inwards due to adhesion between the water and the walls. The counter pull exerted by the walls on the water results in throwing it into a state of tension. Under such conditions the wall pressure and hence the turgor pressure of the cell are negative in value,and the tension (negative pressure) developed within the water will be equal in magnitude to the negative wall pressure. The equation for the diffusion pressure deficit of a cell under such conditions becomes:Diffusion pressure deficit Osmotic pressure (Wall pressure)In other words the diffusion pressure deficit of the water in a cell or vessel when subjected to a tension is equal to the osmotic pressure plus the tension imposed on the water.

The diffusion pressure defi... may
disregard for the time being the ... that cell
range in value from zero to th... its in plant
when flaccid. The range of di... the range of
tissues therefore correspo... this chapter.
osmotic pressures, as dis...

DYNAMICS ... INTERCELLULAR MOVEM... WATER IN PLANTS

In order to ... this prt of the discussion changes in the osmotic ... of cells du solely to volume changes will again be disre... ed, as usuall these arenot great enough to modify seriousl any generalized picture of the water relations of plant cell.

Let us imagine a certain cell. Diagram of two adjacent cells (X) to have an osmotic pressure of used in explanation of the mechanism of 12 atmos., and a wall pressure of cell-to-cell movement of water.6 atmos.; and a second cell (Z) to have an osmotic pressure of 10 atmos., and a wall presure of 2 atmos. Let us further suppose that these two cells are brought into such intimate contact that a normal osmotic movement of water can occur between them, under conditions that no evapouration can occur from them.

Which cell will gain water under these conditions? An uncritical interpretation would answer that water will move from cell Z to cell K, since the cell sap of X has the higher osmotic pressure. Although such an explanation is commonly offered, even in some textbooks, a little consideration shows that this cannot possibly be true.The water in the cell sap of X would have a diffusion pressure deficit of 12 atmos. were it under no pressure. The wall pressure of 6 atmos., however, reduces the diffusion pressure deficit of the cell to 6 atmos. Similarly the water in the vacuole of cell Z would have a diffusion pressure deficit of 10 atmos. were it not also influenced by a wall pressure of 2 atmos. Hence the diffusion pressure deficit of cell Z is 8 atmos. The diffusion pressure deficit of cell X is therefore less than that of cell Z. Water wil there fore move from cell X to cell Z. Water will continue to show a net movement from X to Z until the diffusion pressure deficits of the two cells are equal. In the movement of water from cell to cell in plants it is the diffusion pressure deficits and not the osmotic pressures which tend to equilibrate. This is only a special aspect of the fundamental tendency of the diffusion pressure of water to attain a uniform value throughout any system. It is by no means impossible, therefore, for water to move from a cell of higher to one of lower osmotic pressure, although it is not to be inferred that this is generally or even usually true. Any cell will absorb water from a cell or solution of lower diffusion pressure deficit and, correspondingly, will lose water to a cell or solution of higher diffusion pressure deficit.

It should not be supposed that after a dynamic equilibrium has been established between two cells that their diffusion pressure deficits will be an exact average of their initial diffusion pressure deficits, as this is seldom true. Differences in the original volume of the cells and the elasticity of the cell walls usually make this impossible. The only general statement which can be made is bat, at equilibrium, the diffusion pressure deficits of the two cells will be equal, and this value will lie some where between the two original values of the cells.

Whenever the diffusion pressure deficits of two adjacent cells are dissimilar we may speak of a diffusion pressure deficit gradient as existing between them. Movement of water from one cell to another can occur only when such a gradient exists. Other conditions being equal the steeper this gradient, i.e. the greater the difference in diffusion pressure deficits, the more rapidly one cell will gain water from the other. The term diffusion pressure deficit gradient can also be applied to a chain of cells in which the diffusion pressure deficit increases serially from cell to cell. Several examples of such gradients will be discussed in subsequent chapters.

WATER RELATIONS WITHIN INDIVIDUAL PLANT CELLS

Imbibition of water is due fundamentally to a lesser diffusion pressure of water in the imbibing substance than in the surrounding water. We may therefore also speak of the diffusion pressure deficit of an imbibing substance. A block of dry gelatin, for example, dropped into a solution with an osmotic pressure of 10 atmos. will, after a dynamic equilibrium has been established, have a diffusion pressure deficit of 10 atmos. A completely saturated imbibant will have a diffusion pressure deficit of zero. The development of what we have just called the diffusion pressure deficit of an imbibing substance is a more complicated phenomenon than the development of the diffusion pressure deficit of a solution, but further details cannot be considered here.

Within a plant cell the cell sap, the water in the protoplasm, and the water in the cell wall may each be regarded as possessing a diffusion pressure deficit of its own. The diffusion pressure deficit of the protoplasm and the cell wall are at least in part of imbibitional origin. In an isolated plant cell from which evapouration of water is prevented the diffusion pressure deficits of all parts of a cell are usually approximately in equilibrium. If immersed in asolution all parts of a cell will more or less rapidly attain an equilibrium, not only with each other but with the surrounding solution.

If the magnitude of the diffusion pressure deficit of any part of a cell is either increased or decreased the values for other parts of the cell tend towards equilibrium with this new value. For example, if the diffusion pressure deficit of the mesophyll walls of a cell is increased due to evapouration from them during transpirational water loss, water will diffuse toward the walls from the protoplasm and cell sap as long as their diffusion pressure deficit is less than that of the walls. Similarly if the osmotic pressure of the cell sap is increased by a conversion of insoluble starch to sugars, water will diffuse into the cell sap from the wall and protoplasm until a diffusion pressure deficit equilibrium is attained within the cell.

METHODS OF MEASURING THE DIFFUSION PRESSURE DEFICIT OF PLANT CELLSAND TISSUES

By the methods at present in use it is possible to measure the diffusion pressure deficit only for cells with elastic walls. Most methods of measuring this quantity are based on the principle, previously discussed,that if a cell is immersed in a solution with an osmotic pressure equal to the diffusion pressure deficit of the cell, a dynamic equilibrium is immediately established, and no change will occur in the volume of the cell. In solutions with a higher osmotic pressure than the diffusion pressure deficit of thecell, it will decrease in volume; in solutions of lower osmotic pressure than its diffusion pressure deficit the cell will increase in volume.

The first step in making determinations of the diffusion pressure deficits of plant cells is to prepare a series of sucrose solutions graded at appropriate intervals in osmotic pressure. The range of the series to be used will depend upon the possible diffusion pressure deficits in the material under investigation. Some investigators have undertaken to measure the diffusion pressure deficit of individual cells. Single cells or groups of such cells are isolated from the tissue, and mounted under the microscope usually in paraffin oil (to prevent changes in diffusion pressure deficit due to evapouration, or osmotic movement of water into or out of the cells from or into an aqueous medium). The linear dimensions of the cells are recorded after which a sample of the cells is immersed in each sucrose solution in the series. After the attainment of an equilibrium between the cells and the solutions they are remeasured under the microscope. The osmotic pressure of the solution in which the cells show no change in dimensions is considered to be equal to the diffusion pressure deficit of the cell. This method requires technique of considerable precision and has not been widely used.A simplified method of measuring diffusion pressure deficits has also been introduced by Ursprung and Blum. In this method narrow strips of tissue are cut from such structures as thin leaves or petals. The length of each strip is immediately measured under the microscope, usually while mounted in paraffin oil. Several strips are then immersed in each of a graded series of sucrose solutions in which they are allowed to remain until an equilibrium has been attained, after which they are remeasured. The osmotic pressure of the solution in which no change in the length of the strip soccurs is considered to be equal to the average diffusion pressure deficit of the cells in the strip.

For the measurement of the average diffusion pressure deficit of the cells in bulky tissues such as potato tubers, masses of tissue such as cylinders of approximately equal dimensions can be employed. If possible all of the cylinders used in a given determination should be cut from the same organ. The equilibrium point can be determined by measuring changes either in the weight or the volume of the cylinders. The solution in which the cylinder neither gains nor loses weight (or volume) is considered to have an osmotic pressure equal to the average diffusion pressure deficit of the cells in the cylinder. Determinations of diffusion pressure deficits by the methods described above should not be confused with plasmolytic determinations of the osmotic pressures of plant cells. In the latter type of determination the critical measurement is the osmotic pressure of the solution with which the cells come to an equilibrium with out any turgor pressure, i.e. at incipient plasmolysis. In diffusion pressure deficit determinations the critical measurement is the osmotic pressure of the solution with which the cells come to equilibrium without any change in their turgor pressure, i.e. without any change in the volume of the cells. Only when the cell is initially in a completely flaccid condition will determinations of its osmotic pressure and its diffusion pressure deficit result in the same values.

All of the methods of measuring the diffusion pressure deficits of plant cells are subject to most of the flaws which are also inherent in plasmolytic determinations of osmotic pressure, as well as some other serious errors. Except possibly for a few types of cells which are especially well adapted to such measurements, it is doubtful if the values obtained in determinations of diffusion pressure deficits are ever more than fair approximations of the true values which obtain in plant cells.Neither the turgor pressure nor the wall pressure of a cell can be determined directly. However, if the osmotic pressure of a cell or group of cells has been determined, and Ekewise the diffusion pressure deficit, the turgor pressure or wall pressure can be calculated from the equations given earlier in the chapter.

Chapter 12

The Loss of Water from Plants

It is common place knowledge that all plants require water for their existence and development and that most plants require it in considerable quantities. It is not so generally recognized, however, that in most species of plants an overwhelmingly large proportion of the water absorbed from the soil is lost by the plant into the atmosphere and takes no permanent part in its development or in its metabolic processes. The lack of this general realization is probably due to the fact that while water is supplied to and absorbed by plants in its familiar liquid form, by far the greater part of that lost escapes in the invisible form of water-vapour. The loss of water-vapour from living plants is known as transpiration. Loss of water-vapour may take place from any part of a plant which is exposed to the air. This applies even to roots in contact with the soil atmosphere. Generally speaking, however, the leaves are the principal organs of transpiration. Most of the transpiration from leaves occurs through the stomates;this is termed stomatal transpiration. Smaller amounts of water-vapour are lost by direct evapouration from the epidermal cells through the cuticle; this is usually called cuticular transpiration. All aerial parts of plants lose some water by transpiration, although, due to the presence on some organs of superficial layers almost impervious to water, the rate of loss from most such organs is very low. Some of the transpiration from herbaceous stems, flowerparts, fruits, etc. is of the cuticular type, but is small in amount. Most herbaceous stems, fruits, and flower parts bear stomates which permit the occurrence of stomatal transpiration from such organs. Loss of water-vapour also takes place through the lenticels of woody stems; this is usually called lenticular transpiration.

THE MECHANICS OF FOLIAR TRANSPIRATION

The subsequent discussionwill deal almost entirely with transpiration from leaves, since, in most plants,the amount of water-vapour lost from other organs is comparatively small. The mechanics of foliar transpiration can be adequately discussed only inreference to the anatomy of the leaves from which it occurs. It should be recalled that the vacuoles of all of the living cells of a leaf are filled with water, which also saturates the protoplasm and the cell walls, this water being supplied to the leaf cells through the water conducting tissues of the vascular bundles. Hence evapouration of water will occur from these wet cell walls into the internal atmosphere of the intercellular spaces just as it will occur from any wet surface in-to the surrounding air. The intercellular spaces constitute a connected system, ramifying throughout the leaf.

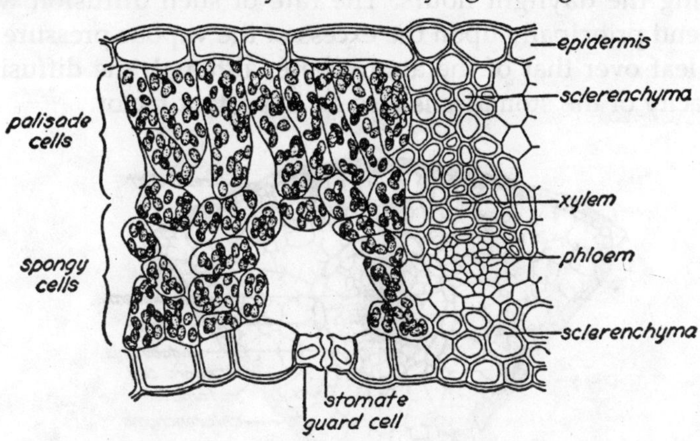

Fig. Cross section of a portion of a leaf of tulip tree

If the stomates are closed the only effect of evapouration from the mesophyll cell walls will be the saturation of the entire volume of the intercellular spaces with water-vapour.

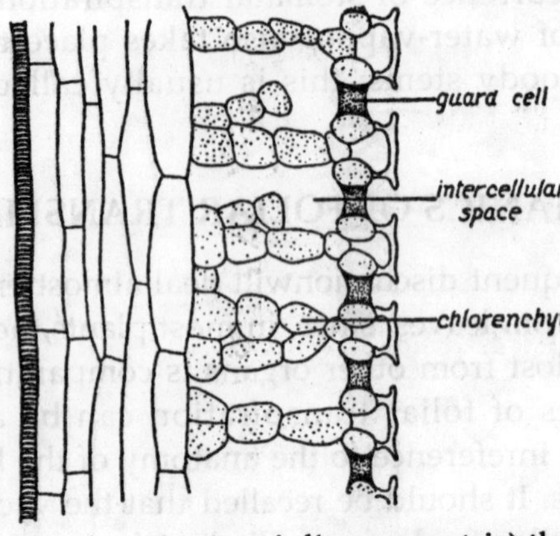

Fig. Longitudinal section(semi-diagrammatric) through a small portion of a leaf of white pine(pinus strobus).

When the stomates are open, however, diffusion of water-vapour may occur through them into the outside atmosphere. Such outward diffusion will always take place unless the atmosphere has a vapour pressure equal to or greater than that of the small portion of a leaf which does not commonly exist during the daylight hours. The rate of such diffusion will depend principally upon the excess of the vapour pressure in the leaf over that of the atmosphere, although the diffusive capacity of the stomates is also an important factor.

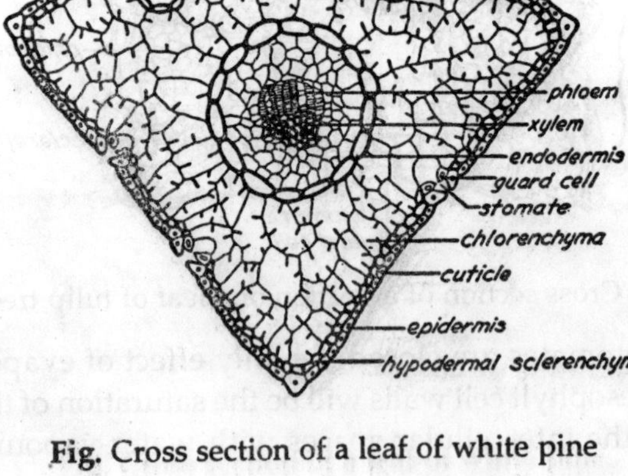

Fig. Cross section of a leaf of white pine

The process of stomatal transpiration therefore involves evapouration from the cell wall surfaces bounding the inter cellular spaces and the diffusion of this water-vapour from the intercellular spaces into the atmosphere through the stomates. One side of every epidermal cell on a leaf is also exposed to the atmosphere. Evapouration of water occurs into the atmosphere directly from these cell surfaces. The surfaces of practically all aerial leaves are covered with a layer of wax-like substance known as cutin reduces transpiration directly through the walls of the epidermal cells to a magnitude much less than it would be, were there no such layer present. The thickness of the cutin layer varies with the species of the plant, and the environmental conditions under which the leaves have developed. The layer of cutin is usually thicker on leaves which have developed in bright sunlight for example, than on leaves of the same species which have developed in shade. Even in leaves which are heavily coated with cutin, some cuticular transpiration occurs, possibly largely through tiny rifts in the cutin layer. In most species of plants of the temperate zone less than 10 per cent of the foliar transpiration occurs through the cuticle, the remainder being stomatal transpiration.

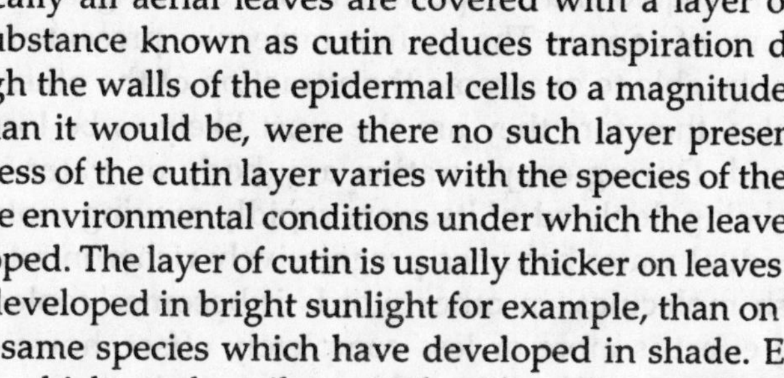

Fig. Citin layer on the upper epidermis of a leaf of clivia nobilis.

EVAPOURATION

The fact that transpiration may be regarded as essentially a modified form of the process of evapouration makes desirable a fuller consideration of the dynamics of this process. When an open pan of water is exposed to the atmosphere, the level of the water in the pan slowly drops. Evidently water molecules are being slowly lost into the atmosphere. This is an example of the well known process of evapouration. All of the molecules in a mass of water are not travelling with the same velocity, although for any given temperature the average speed of all of the molecules in the liquid mass is a constant. Some of the molecules in the liquid water attain sufficient momentum to entirely overcome the attractive forces holding them in the liquid and escape into the surrounding atmosphere in the form of vapour. The swiftest molecules present are most likely to be able to overcome the attraction of the other water molecules, therefore they are the most likely to be lost from the liquid. During evapouration any body of water is thus slowly being depleted of its more rapidly moving molecules. The residual water becomes progressively richer in relatively sluggish molecules; in other words it becomes cooler. This cooling effect is more or less completely offset, however, by physical transfer of heat into the water from its surroundings as soon as its temperature drops below that of the environment. The proportion of high speed molecules in the molecular population of the pan of water is thus maintained at close to its original value and the rate of evapouration continues with very little diminution.

In the preceding paragraph we have focussed our attention only on thewater molecules which escape from the liquid. Water-vapour molecules are also returning to the liquid from the atmosphere during the process of evapouration. If a water-vapour molecule in the atmosphere above the evapourating surface, particularly one of the more sluggish ones, strikes the surface of the water at the proper angle and with not too great a velocity, it will be held there by the attractive forces exerted by the liquid water molecules, and again become part of the liquid water. High speed molecules, on the other hand are much less likely to be captured by the attractive forces exerted by the molecules at the surface of the body of liquid. They impinge upon the bounding film of the liquid with such velocity that unless they hit that surface a stright angles or nearly so, they usually glance off along a new pathway.This picture of the kinetics of the evapouration process holds for any evapourating surface whether it be the exposed surface of a body of water a moist piece of cloth, paper, or porous clay, or the mesophyll cell walls of a leaf. Ordinarily the term evapouration is used to refer only to conditions in which the rate of escape of molecules exceeds their rate of return, and its use will be restricted to this sense in this discussion.

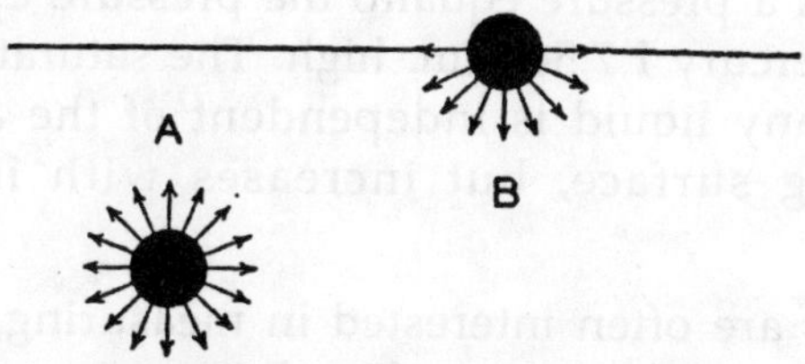

Fig. A diagrammatic representation of the direction of the forces of cohesion acting upon a water molecule: (A) in the body of the liquid, (B) at its surface

If the air above the water surface is confined, as for example, when a dish of water is covered by a bell jar, the number of water-vapour molecules in the confined space will gradually increase due to evapouration from the surface of the water. Since their movement is a random one the water-vapour molecules will be continually colliding with walls of the container, each other the molecules of other gases present in the air, and surface of the liquid water. Some of those which strike the surface of the liquid will be held there by intermolecular attractive forces. As the concentration of water-vapour molecules in the air increases, the number which plunge back into the water in any unit interval of time will increase. Eventually a dynamic equilibrium will be attained at which the number of molecules leaving the surface and the number returning to it in a unit time will be equal. At this point the airwill be saturated with water-vapour and evapouration will no longer be occurring.

The water-vapour molecules exert a definite pressure against the walls ofthe container and the surface of the water. This is known as vapour pressure.In the illustration given in the preceding paragraph, tihe vapour pressure of the water increases progressively until the saturation point is reached. The vapour pressure of water under the conditions of such a dynamic equilibriummay conveniently be termed the saturation vapour pressure. The vapour pressure of a liquid is usually expressed in terms of millimeters of mercury. The saturation vapour pressure of water at 2O C., for example, is I 7.54 mm. Hg.This means that at this temperature the water-vapour exerts a pressure equalto the pressure exerted by a column of mercury I 7.54 mm. high. The saturation vapour pressure of any liquid is independent of the area of the evapourating surface, but increases with increase in temperature.

Botanists are often interested in measuring the rate of evapouration undera given set of conditions in connection with studies of transpiration and of plant distribution. The difficulties involved in making this apparently simple measurement are not all apparent at first consideration. It would appear that the rate of evapouration could be measured by the rate at which water disappeared from an open pan exposed to the atmosphere. While such methods have been used they are subject to numerous errors and limitations.The rate of evapouration for such a pan is controlled not only by environmental conditions, but also by the size, shape, and colour of the pan, and the depth of water below the rim. Other errors arise from the accumulation of rain water and the splashing of water due to wind or other causes. The many difficulties encountered in the use of open pans of water for measuring evapouration rates led to the devising of more suitable instruments for this purpose. Many measurements of evapouration rates have been made with instruments called atmometers. These consist essentially of surfaces of porous clay moulded in the form of hollow cylinders or spheres.Water evapourates from such surfaces in essentially the same way that it evapourates from a free surface of water. Atmometers are attached to a reservoir of water and are usually provided with mercury valves which prevent absorption of rain. The loss of water from such instruments can be determined either by measuring the decrease in volume of water or the loss in weight of the instrument. Although atmometers are not subject to many of the errors inherent in the use of open-panevapourimeters numerous precautions must be observed in employing them. Details of atmometric technique are discussed by Livingston Measurement of Transpiration. Four general methods are in use for measuring transpiration, but only the first two of these as listed can be used for quantitative determinations of the rate of loss of water-vapour from plants which are rooted in the soil.

METHOD OF WEIGHING POTTED PLANTS

This method can be employed only with plants which are rooted in pots or other suitable containers. For laboratory experiments potted plants are often used, the pot generally being enclosed in a metal shell, and the soil surface sealed off so that evapourational water loss can occur only through the plant. For field or large-scale experiments, metal receptacles have been found convenient in which case it is only necessary to seal off or cover the soil surface in such a way as to prevent evapouration. The method is limited in practical application to plants which can be grown in readily portable containers. The transpiration rates of plants as large as mature maize plants have been measured by this method. The loss in weight of the container and plants for a given time interval may be considered to represent transpiration as the effect of other factors on the weight of the set-up is usually negligible. If the experiment is to be continued for more than a relatively short period it is necessary to provide the container with a watering tube through which known volumes of water can be introduced into and distributed throughout the enclosed soil at appropriate intervals.

In employing this method the receptacle in which the plants are growing may be weighed at selected intervals by manual manipulation or may be placed on a balance which is so arranged that each loss of a definite increment of weight (for example i g.) is automatically registered on a recording device.

METHOD OF COLLECTING AND WEIGHING TRANSPIRED WATER-VAPOR

This method requires a rather elaborate experimental set-up, but is the only way in which the rate of transpiration can be determined quantitatively for plants growing in out-of-door habitats. The plant is enclosed within a glass chamber, the soil surface sealed off, and a stream of air taken directly from the atmosphere circulated rapidly through this container. After passing through the chamber the air is conducted through tubes or vessels containing a water absorbent such as calcium chloride. The gain in weight of these absorption tubes during the experiment represents the water-vapour transpired by the plant plus the water-vapour which was introduced into the system from the outside atmosphere. In order to determine the proportion of the water-vapour which comes from the atmosphere it is necessary to set up a check apparatus in exactly the same manner, except that no plant is enclosed in the glass chamber, and to pass air through it at the same rate that it is circulated through the set-up containing the plant. The gain in weight of the water-absorption tubes in the second apparatus represents water-vapour from the atmosphere. Such a method has been used by Minckler for measuring transpiration from attached branches of trees.

POTOMETER METHODS

Of a limited usefulness for the measurement of transpirational water loss are instruments known as potometers. The severed stem of a plant is immersed in water in the reservoir of the potometer and the rate of water loss determined by the rate at which the volume of water in the apparatus shrinks. This is usually followed by noting the rate of movement of an air bubble in the water in the capillary side arm of the instrument. Some potometers are so constructed that the entire root system of plants which have been specially grown for the purpose in solution cultures can be immersed in the reservoir of the instrument. A potometer actually measures the rate of absorption of water rather than the rate of transpiration. While under many conditions the rates of these two processes are virtually equal, this is not always true, particularly if an internal water deficit exists in the plant. The rate of transpiration as measured for a cut shoot in a potometer does not necessarily bear any relation to its rate of transpiration while it was still attached to a plant. The reasons for this will become clear in the discussion of the internal water relations of plants. The principal utility of potometers is in laboratory demonstrations of the effects of various environmental factors upon the rate of transpiration.

HYGROMETRIC PAPER METLZODS

When filter paper is impregnated with a dilute (about 3 per cent) solution of cobalt chloride and dried it becomes a bright blue in colour. If exposed to moist air its colour gradually changes to pink. The same colour change ensues when a piece of such paper is brought into contact with the transpiring surface of a leaf. If small pieces of such paper are so mounted as to be protected from the water-vapour of the atmosphere by glass, mica, or celluloid, and are brought into contact with the surface of a leaf the rate at which the paper changes in colour from blue to pinkish an indication of the rate at which water-vapour is being lost by that leaf. A leaf which changes the colour of a piece of cobalt chloride from its full blue colour to its full pink colour in, for example, 30 seconds is losing water to the paper twice as fast as a leaf which requires 60 seconds to accomplish a similar change. This method gives no measure of absolute rates of transpiration because when a portion of a leaf is covered with a piece of cobalt chloride paper the environmental conditions influencing the leaf under the paper are very different from those which would influence it were it freely exposed to the atmosphere. The leaf under the paper is exposed to a edietd Iight Intensity and an initially lower vapour pressure than freely exposed parts of the same leaf, and furthermore, is completely isolated from any effects of wind. Hence the rate of loss of water to the paper may be very different from the rate of water loss of the same area of leaf surface to the atmosphere. Under certain conditions this method can be used for a determination of the relative rates of transpiration of different species with a fair degree of accuracy. Even for relative determinations of transpiration rates, however, this method gives valid results only when all of the plants are growing under essentially the same atmospheric conditions. Modifications and limitations of this method are discussed by Livingston and Shreve and Meyer.The transpiration rates of plants may be expressed in terms of a unit of leaf surface (usually per square decimeter), per unit of fresh or dry weight of the leaves or tops of the plant, or per plant. Each of these modes of expression of transpiration rates is of value for certain purposes.

THE MAGNITUDE OF TRANSPIRATION

The Magnitude of Transpiration whether computed on the basis of a unit area of leaf surface, an entire plant or an acre of forest or crop plants, shows a great variation depending both on the plant species and environmental conditions to which they are exposed. The quantity of water transpired may be computed and compared for hourly daily, seasonal, or yearly periods. Because of the great fluctuations in transpiration rates with environmental conditions the citation of isolated values of transpiration rates is usually of very little physiological significance.

It is impossible to calculate accurately the transpiration rate of large plants such as trees from the known rate of transpiration of some of the leaves and an estimate of the area of all of the leaves on the tree. In the first place it is difficult to arrive at any accurate estimate of the aggregate leaf area of a large tree. In the second, there is a great variability in the transpiration rate among the different leaves. The rate of transpiration of the individual leaves varies depending upon their position, exposure, age, and internal physiological conditions. For example, Huber found that low twigs of a redwood tree transpired six times as rapidly as similar twigs from the height of 12 meters. Neither can it be assumed, as mentioned in the discussion of the potometer method, that the transpiration rate of leaves of branches which may have been removed from a tree and placed with their cut ends in water will be the same as their rate of transpiration before the branches were detached.

Hence attempts to determine the transpiration rate for an entire tree from values determined for a few of the leaves on that tree can only result inapproximations. The same comments in general apply, although perhaps not so emphatically, to calculations of the total transpiration of fields of cultivated crops and natural vegetation areas based upon transpiration as determined for a limited number of individual plants.

If transpiration is measured for one minute intervals rates much in excess of this may be observed, but such rates are maintained only for relatively short periods of time of necessity approximations, although, not of course, without their utility in the consideration of certain types of problems. The results of such calculations indicate that sufficient water may transpire from maize plants during the course of a season to cover the field in which they are growing to a depth of 15 inches. Similarly Minckler has calculated that in a growing season water equivalent to 4.8 inches of rainfall may transpire from an acre of American elm trees and that an acre of red maples, growing in a very moist habitat, may lose water equivalent to 28.3 inches per acre. In evaluating all such data it must be remembered, however, that the magnitude of transpiration varies tremendously with the available soil water supply.

SIGNIFICANCE OF TRANSPIRATION

Opinions regarding the significance of transpiration have ranged all the way from those which would put it practically on a par with such processes as photosynthesis and respiration, to those which would relegate it to the category of a necessary evil. The principal roles which have been ascribed to the transpiration of plants can be summarized under the following three headings:

Supposed Role in the Movement of Water.

It is often claimed or assumed that the movement of water through the plant requires the occurrence of transpiration from the leaves. That this concept is entirely erroneous will be clear from the discussion of the mechanism of the conduction of water through plants. Under conditions of high transpiration the movement of water through plants is, in general, more rapid than under conditions of low transpiration. The mechanism causing ascent of water through a plant operates in such a way that any decrease in the turgidity of the mesophyll cells favours a more rapid movement of water towards those cells. Hence a rapid transpiration rate, which invariably results in a considerable loss of turgidity by the mesophyll cells, usually speeds up the rate at which water ascends through the plant. However translocation of water to the extent that it is used in photosynthesis, growth or other metabolic processes continues even if the transpiration rate is negligible.

Supposed Role in the Absorption and Translocation of Mineral Salts.

It has often been assumed that the more rapid the rate of transpiration, the greater the rate of absorption of mineral salts. This view implies that the dissolved mineral salts are swept into the plant along with the absorbed water, a postulation which ignores much evidence that the mechanisms operating in the absorption of water and the absorption of mineral salts are very different.

The results of certain experiments do indicate that a somewhat larger quantity of mineral salts accumulates in plants under conditions favouring high transpiration than in similar plants growing under conditions of low transpiration although there is no actual proportionality between the volume of water absorbed and the quantity of mineral salts which pass into the plant. Since even when plants grow under conditions favouring low transpiration rates they usually obtain an adequate quota of the various mineral salts, provided they are present in sufficient abundance in the soil, it is difficult to see that this effect is of any very great advantage to the plant.

Realization is now fairly general that there is, at the most, only a slight correlation between the rate of transpiration and the rate of absorption of mineral salts from the soil.

It is also generally considered that transpiration plays an important part in the translocation of mineral salts from the root system to the top of the plant. That some upward translocation of mineral salts occurs in the transpiration stream is indisputable, but whether this is an important part of the total quantity transported is open to debate. There is considerable evidence that at least a part of the mineral salts which move through the plant are translocated in the phloem. The rate of transpiration can have little or no direct effect upon the movement of such salts. While high transpiration rates may mean a somewhat higher rate of transport of mineral salts through the plant than low ones it is doubtful if this is of great significance from the standpoint of the plant.

Supposed Role in the Dissipation of Radiant Energy.

Leaves exposed to direct sunlight absorb large quantities of radiant energy which, unless dissipated in some other way, will be converted into heat energy and raise the temperature of the leaf. The possibilities of such an effect are indicated by the following approximate calculations: In direct noon-day summer sunlight the rate of receipt of solar energy is often 1.3 g.-cal. per square centimeter of leaf surface per minute and not uncommonly even greater. The proportion of this actually absorbed varies with the kind of leaf, but will be assumed to be only 50 per cent, which is a conservative estimate. The incident radiant energy which is not absorbed by the leaf is all transmitted or reflected. Such a small proportion of the absorbed energy is used in photosynthesis that it can be neglected in a rough calculation. Hence about 0.65 g.-cal. of energy is absorbed per square centimeter of leaf surface per minute. If the mass of a square centimeter of a leaf is taken as 0.020 g. and its specific heat as 0.879g.–cal.2 the rise in temperature per minute would be 0.60.020 × 0.879 or about 370 C. Since the thermal death point of most plant protoplasm lies. These are the actual values as determined for a sunflower leaf by Brown and Escombe.

It should not require more than two minutes to heat the leaves of most plants to a lethal temperature. Actual measurements of leaf temperatures, however, show that they seldom even approach their thermal death points. Leaf temperatures usually do not exceed atmospheric temperatures by more than a few degrees centigrade. Evidently some efficient energy-dissipating mechanism is at work which prevents the accumulation of heat energy in leaves.

Since transpiration is an energy-consuming process it has often been assumed that it is in the evapouration of water from the leaves that most of the energy absorbed by them is dissipated. We might well inquire, therefore regarding the possible efficiency of transpiration as an energy-dissipating process. The evapouration of a gram of water at 20 C. requires 584 g.-cal. For the dissipation of 0.6 g.-cal. of heat, therefore, the evapouration of 0.0011 g.(0.65) of water per square centimeter of leaf surface per minute is required. This is equivalent to 6.6 g. (0.0011 × 100 × 60) of water per square decimeter of leaf surface per hour, a rate of transpiration which is seldom if ever attained by plants for any sustained period of time. Evidently transpiration even when occurring at its maximum rate, can at the most account for a dissipation of only part of the radiant energy which is absorbed by leaves indirect sunlight.

The fact that transpiration is often inadequate in direct sunlight to account for the dissipation of all the absorbed radiant energy leads naturally to the question of whether it is at all essential for this process. Observations have shown that leaves in which the transpiration rate is greatly reduced, as for example those in which the stomates are plugged with vaseline, leaves in a wilted condition, or the leaves of plants in xeric habitats during dry seasons in which the occurrence of any appreciable amount of transpiration is precluded by the lack of soil water, seldom have temperatures which are any where near the thermal death point, even when exposed to direct sunlight. Although transpiration often accounts for the dissipation of some or even most of the absorbed radiation, in so doing it apparently plays no essential role since absorbed radiation can be dissipated by purely physical means. As soon as the temperature of a leaf exceeds that of the surrounding atmosphere, it will lose heat to the atmosphere in the same way that any other object heated above its environmental temperature does, i.e. by one or more of the purely physical processes of conduction, radiation, and convection. The term thermalemissivity is frequently applied to this physical loss of heat energy by leaves and other objects.

Actually, instead of benefiting plants, transpiration may often be detrimental. Under conditions of deficient soil water or of high transpiration rates even when the soil water supply is adequate, this process results in adiminution in the water content of a plant and in the turgidity of its cells. Prolonged drought conditions ultimately result in a severe desiccation withthe consequent death of all except the most drought resistant species. When the diminution in water content is less severe, a train of other effects such as a decrease in the turgidity of the cells, stomatal closure and reduction in rate or cessation of photosynthesis are induced, all of which have the end result of checking the growth of the plant. It is probably true that lack of water is more often the limiting factor in plant growth than any other single factor. Furthermore deficiency of water is probably responsible for the death of more plants under natural or even cultural conditions than any other single cause.

The fundamental effects of transpiration upon the plant are not to besought in any hypothetical advantages of the process to the plant, but in its very real and readily ascertainable influences upon the water relations of plant cells and tissues, and through these its effects upon other plant processes. In spite of the fact that transpiration may be regarded in a sense as an incidental phenomenon, its indirect influence upon the metabolic processes of plants is a profound one. It is this fact, more than any other, which justifies intensive and critical studies of this process.

THE LOSS OF WATER FROM PLANTS IN LIQUID FORM

If a pot of young oat plants be copiously watered and then enclosed in a bell jar, in a relatively short time a slow exudation of water begins at the tip of each leaf. The drops which form at the leaf apices gradually enlarge and eventually may run down the side of the leaf or fall off. This process of the escape of liquid water from uninjured plants is called guttation. It is of very general occurrence in plants, baying been recorded in plants of more than 300 genera although there are many species in which it has not been observed. Guttation occurs most frequently and abundantly under conditions which favour rapid absorption of water by the roots but which result in a reduced rate of transpiration. In most temperate regions such conditions occur most frequently during the late spring when there is often an alternation between relatively cool nights and relatively warm days. Guttation is frequently observed at that season, usually taking place at night or during the early hours of the morning. The drops of guttation water which form at the tips of grass blades and the tips or edges of the leaves of other plants are often erroneously considered to be dew.

The process of guttation occurs from specialized structures known as hydathodes. These structures are also sometimes referred to as water stomates or water pores. A typical hydathode consists of an enlarged stomate-like opening below which is a rather large. Guttation from tomato leaves. Photograph, courtesy of Dr. J. H. Gourley chamber, bordered by a mass of thin-walled, loosely arranged cells called the epithem. The xylem elements of a vascular bundle terminate just below the epithem.

The exudation of water through hydathodes is considered to result from develops in the 'ap of the xylem elements, and not to any locally developed pressure in the hydathode itself. This pressure is generally believed to be identical with the so-called root pressure. It is supposed that the water is forced from the vessels through the intercellular spaces of the epithem layer and out of the plant through the pore of the hydathode. This water is not pure but contains traces of sugars and other solutes. While the volume of water exuded by most temperate region plants in guttation is usually insignificant, some tropical species lose large quantities in this process. A young leaf of Colacasia nymphaefolia, a native of India, has been observed to lose as much as 100 cc. of liquid water in a single night by guttation.

Glands are found on leaves, flower parts and other organs of the plant. Many of them represent modified epidermal hairs or other epidermal cells. Certain types of glands secrete water or, more accurately, a dilute sap. These have sometimes been classed as hydathodes, but it seems better to restrict this term to the type of structure previously described under this name epithem. The exudation of water by glands is generally called secretion and is apparently caused by forces which develop within the gland itself, and not by a pressure developed in the sap of the water conducting tissue as appears to be true of hydathodes. Glands which secrete water or dilute saps are not closely connected with xylem elements as are the hydathodes which secrete other substances as well as water. Among these are sugars (as in nectar), calcium salts, resins, volatile oils, and enzymes.

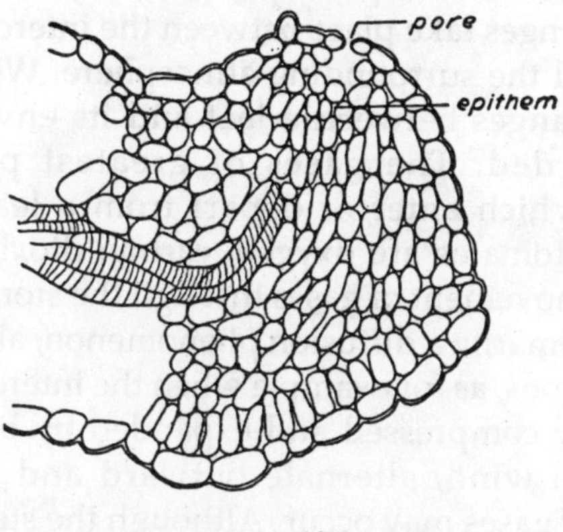

Fig. Hydathode at the margin of a tomato leaf as seen in sectional view (semi-diagrammatic). Note termination of vessel just back of the epiterm.

If the stem of an herbaceous plant be cut or broken a slow exudation of sap often occurs from the ruptured stump. A similar phenomenon can often be observed in the cut or broken stems or branches of woody plants, especially in the spring of the year. The exudation of sap from holes bored into maple trees is a good example. This phenomenon has long been called bleeding. Under certain conditions large quantities of dilute sap may be lost in this process. A single vigorous grape vine may lose as much as one liter of sapin a day, while under favourable conditions a sugar maple tree will yield from five to six liters in twenty-four hours. In some species bleeding appears to be due to a root pressure developed in the xylem elements, in others to pressures developed in the phloem, and instill others to locally developed pressures in other tissues. The significance of the various internal pressures which develop in the conductive tissues of plants will be discussed later, particularly in relation to the problems of the translocation of water and solutes in plants.

THE STOMATAL MECHANISM

The most important physiological fact about the stomates is that they are sometimes open and sometimes closed. When open they serve as the principal pathways through which gaseous exchanges take place between the intercellular spaces of the leaf and the surrounding atmosphere. When closed all gaseous exchanges between a leaf and its environment are greatly retarded. The gases of greatest physiological importance which enter or depart from a leaf principally through the stomates are oxygen, carbon dioxide and water vapour. The movement of gases through the stomates in either direction is primarily a diffusion phenomenon, although under certain conditions, as for example when the intercellular spaces are alternately compressed and expanded by bending of the leaf in a high wind, alternate outward and inward mass movements of gases may occur. Although the stomates are the principal portals through which entry and escape of gases takes place, the fact shouldn't be overlooked that at least small amounts of these three gases pass directly through the epidermis and cuticular layers of all leaves. In submerged aquatics all gaseous exchanges between the plant and its environment occur through the epidermis.

STRUCTURE OF THE STOMATES

The stomates or stomata (singular stomate or stoma) are minute pores which occur in the epidermis of plants. They are surrounded by two specialized epidermal cells known as the guard cells. Stomates may occur on any part of a plant except the roots, but in most species are most abundant upon the leaves. The size of the stomatal pore varies in most plants depending upon the turgidity of the guard cells and often, especially at night, it is entirely closed. The structure of stomatal apparatus shows marked variations in detail indifferent species of plants, but the essential feature of a pre between two modified epidermal cells is common to all species of vascular plants. Guard cells which are roughly kidney or bean-shaped as seen in surface view are typical of more species of plants than any other kind. In some species the epidermal cells bordering on the guard cells are different in con figuration from other cells in the same tissue; these are called subsidiary cells or accessory cells. In many species of the grass and sedge families the guard cells are distinctly elongate. In most species of· conifers, and in certain other species, stomates are of the sunken type

The guard cells differ from the other epidermal cells in that they contain plastids which are green in colour. These have often been assumed to bechloroplasts, and are frequently so-called, but the green pigment present does not seem to be true chlorophyll as photosynthesis apparently does not occurring the guard cells.

SIZE AND DISTRIBUTION OF STOMATES

The size of the stomatal pore varies greatly according to the species of plant, and somewhat even among the individual stomates on any one plant. They are always very minute however, their dimensions being expressed in terms of microns. Minute as these openings appear to be from a human scale of values they are enormous when compared with the size of the gas molecules which diffuse through them. The calculated diameter of a water molecule is 0.000454. More than 2,000 water molecules would have to be placed side by side to measure a distance of 1. The molecules of both carbon dioxide and oxygen are larger than water molecules. Since the stomatal diameters usually are considerably in excess of 1, it is evident that the stomates afford relatively enormous portals to the gas molecules which diffuse through them. In general the number of stomates present in the epidermis of leaves may range from a few thousand to over a hundred thousand per square centimeter the exact number depending upon the species and upon the environmental conditions under which the leaf has developed. A single maize plant has been estimated to bear from 140 to 240 million stomates, while the number on a large tree could be expressed only by a figure of astronomical dimensions. However, marked deviations from such average values are possible for any species, depending upon the environmental conditions under whichthe leaves have developed. The number of stomates per unit area of leaf surface may be quite different on leaves of two plants of the same species if one grew in a greenhouse, and the other grew in the open, or upon the leaves of plants of the same species which have developed during different seasons.

Stomates occur in both the upper and lower epidermis of many species of plants. In numerous others, especially woody species, they are confined to the lower epidermis. Even in those species in which stomates occur on both surfaces of the leaf they are commonly, but not always, more abundant in the lower epidermis. In floating leaves such as those of the water-lily, stomates occur only in the upper epidermis. Species in which the stomates are relatively small usually have more per unit area than species in which the stomates are relatively large.

The number of stomates per unit area usually varies on different leaves of the same plant, and even in different parts of the same leaf. On individual leaves it appears to be generally true that the greatest number of stomates per unit area is at the tip, the lowest towards the base, while the middle portion of a leaf shows a frequency midway between these two extremes. As a rule, the higher the point of attachment of a leaf on the stem of a plant the greater the number of stomates per unit area.

In general no correlation has been found between transpiration rates and either the size or distribution of stomates, other factors being much more important in determining the rate of loss of water-vapour from the intercellular spaces.

PRINCIPLES GOVERNING DIFFUSION OF GASES THROUGH THE STOMATES

Since virtually all gaseous exchanges between the intercellular spaces and the atmosphere take place through the stomates, the problem of the diffusive capacity of the stomata1 pores is an important one. Although the aggregatearea of the fully open stomates on a single leaf is seldom more than 3 percent of the leaf area and is often as low as i per cent, the rate of water loss from the leaves (i.e. loss per unit area in a unit time) in many species maybe, under favourable conditions, as much as 50 per cent of the evapouration from an exposed water surface of comparable dimensions. Stlfelt (1932) records that transpiration from a leaf of birch (Betula pubescens) under the most favourable conditions may be 65 per cent of the evapouration from a comparable evapourating surface, but this is probably an unusually high value. It is therefore evident that water-vapour often diffuses through the stomates at rates ranging up to at least 50 times greater than it diffuses away from an equal area of exposed evapourating surface.

This unexpectedly high diffusive capacity of the stomates is intelligible in terms of the results of studies upon the principles of diffusion through small apertures. The classical investigation of this problem was made by Brown and Escombe who studied the rate of diffusion of carbon dioxide through tubes of known dimensions. They first made the important discovery that if a septum which had been pierced with a small circular aperture was interposed across a column of diffusing gas, that the rate at which carbon dioxide diffused through this aperture was much greater than the rate at which it passed through an equal area of the unobstructed tube.

Since the problem at present under consideration is the diffusion of water-vapour rather than carbon dioxide through small apertures will be used to illustrate more fully the principles regarding the diffusion of gases through small openings. These data were obtained by sealing thin septa, through the centre of which were cut round openings of known dimensions, across the circular mouths of small bottles which had previously been filled to a certain level with water, and then determining the loss in weight of each bottle after all of them had been kept under uniform conditions for the same period of time.

The results of this experiment illustrate two important general principles:

1. The quantities of water-vapour diffusing through small openings in a given period of time are proportional (essentially) to the perimeters (circumferences) and not to the areas of the pores.
2. The smaller the pore, the greater the water loss per unit area. The pore in septum 2, for example, has an exposed area only slightly more than one-third as great as the pore in septum I, but diffusion through it is nearly two-thirds as great as through septum i. Similarly septum 10, with only half the exposed area of septum 9, permits almost as much diffusion to occur as through sputum 9. These results indicate that the rate of diffusion of water-vapour through a small aperture is greater than through an equal area of a largere vapourating surface.

In other experiments, Sayre found that diffusion of water-vapour through small openings of elliptical shape is also more nearly proportional to their perimeters than to their areas.Hence the perimeter law for the diffusion of gases also holds for openings of other than circular shape. An important inference which can be drawn from this finding is that a stomate attains almost its maximum diffusive capacity long before it is fully open, since the perimeter of a stomatal pore does not increase greatly after the aperture between the guard cells has widened perceptibly.

The fact that the diffusion of gases through small apertures is much more nearly proportional to their perimeters than to their areas is due to the more rapid diffusion of the molecules through an opening at the peripherythan through its centre. A molecule of the diffusing gas at the rim of the hole is surrounded on all sides by other molecules moving in all directions.The concentration of diffusing molecules is much less in an outward direction from the rim, however, than towards the centre of the opening. Furthermore, the diffusion gradient is much steeper from the rim outwards than at points above the opening. Hence the number of molecules escaping through the hole from a given point near the rim per unit time interval will greatly exceed the number escaping from a point near the centre of the opening.

Reduction in the area of any circular or elliptical opening results in increasing its perimeter relative to its area. Hence in small apertures such alarge proportion of the diffusion occurs at the edge of the opening that the effect of any diffusion which occurs through its centre is almost if not completely obscured, and measurements show the rate of diffusion through the small apertures to be essentially proportional to their perimeters. In fact in an opening 0.5 cm. in diameter, for example, a large part of the centre of the aperture can be entirely blocked off without any very great effect upon the rate of diffusion of a gas through it.

That the diffusion of gases through small apertures occurs largely at the rim of the hole can be demonstrated visually by means of a simple experiment. A cylindrical glass vessel is partly filled with a gelatin sol which is allowed to gel. A small opening is punched through the centre of a disk of celluloid which is then inserted into the position on top of the gelatin,the gap between the edge of the disk and the wall of the cylinder being sealed with melted vaseline in order to prevent leaks. A solution of potassiumpermanganate, which is purple in colour, is now introduced into the vessel ontop of the disk. After a few hours the molecules of potassiumpermanganate, after diffusing through the aperture, become distributed in such a way as to occupy a hemispherical zone below the opening. The hemispherical distribution assumed by molecules which are diffusing through a small opening is often spoken of as a diffusion shell. Gases also form diffusion shells above small pores but their configuration is often relatively unstable due to their ready distortion by wind or convection currents. Actually when molecules are diffusing through a small orifice there is formed not one, but two diffusion shells, one on each side of the diaphragm which is pierced with the opening. The experiment described above permits discernment of the hemispherical pattern of diffusion only on the receiving side of the system.There is also a second shell, a mirror image of the first adjacent to the aperture on the side of the system from which diffusion is occurring. The deep colour of the permanganate solution prevents visible detection of this second shell in the experiment described above.Diffusion of gases from or into a leaf through the stomates involves a much more complex system than is represented by a septum pierced by a single aperture.

Diffusion is occurring not through a single opening of potassium permanganate upon diffusion. Experiments into gelatin have been performed in analogous physical systems in which multiperforate instead of uniperforate diaphragms have been used.

The diffusion per pore increases with increase in the distance apart of the apertures, although not proportionately. With pores of this diameter (0.3 mm.) nearly the maximum diffusive capacity is attained when they are spaced at intervals of 20 diameters. The closer together the pores in a septum the greater the overlapping of the diffusion shells. The molecules diffusing through each aperture invade in part the zones into whichthe molecules passing through neighbouring pores would diffuse were each opening the only one in the septum. Hence the diffusion gradients are less steep than they would be were there only one pore in the septum. As a result the rate of diffusion through each pore is less than it would be through a pore of equal dimensions in a uniperforate septum. The loss of water-vapour by diffusion per septum decreases with increase in the distance between the openings, i.e. with decrease in the number ofpores per septum. The decrease in diffusive capacity is not however, in proportion to the reduction in the aggregate area of the pores in the septum. For example, when the pores are spaced 5 diameters apart their aggregate area is only 3.38 per cent of the septum area, yet diffusion through them was 62 per cent of that through an open bottle with a mouth of the same area as the septum.

The dimensions of the pores (0.3 mm. diameter, 0.0707 mm^2 area) used in the experiments just discussed are much greater than those of the average stomate.What about the diffusive capacity of still smaller pores? Huber has shown that the smaller and more numerous the pores, aggregate pore area being constant., the greater the diffusion per septum.1 For example,when pores 0.05 mm. in diameter were used, occupying 3.2 per cent of the1 If equal-sized pores are spaced equidistantly, and represent the same proportionate area of the septum, they will be the same distance apart in terms of their diameters, regardless of pore size.

Hubers results indicate that the smaller the pores for a given aggregate pore area the more nearly the diffusion through the septum will approach that of a free evapourating surface. In the sunflower and undoubtedly also in some other species the aggregate area of the stomatal pores is about 3 percent of the leaf area. If diffusion through pores 0.05 mm. in diameter,occupying 3.2 per cent of the area of the septum, is 72 per cent of that from a free evapourating surface, diffusion through the much smaller stomates of this species should be proportionately greater. In such species it seems probable that the aggregate diffusive capacity of the stomates mayclosely approach that of a free evapourating surface. Even in species in which the total area of the stomatal pores is less than in the sunflower their aggregate diffusive capacity is very large relative to their area. The relatively high rates of water loss which occur from leaves in proportion to the area of the stomatal pores are thus explicable in terms of the principles of diffusion through multiperforate septa.

The theoretical diffusive capacity of the stomates is more than adequate to account for known rates of transpiration. This is shown by the calculations of Brown and Escombe for sunflower leaves. Assuming a leaf temperature of 200 C., the intercellular spaces to be saturated with water vapour at that temperature (vapour pressure 17.54 mm. Hg), and the vapour pressure of the atmosphere to be one-fourth of this value, their computations show that transpiration would occur at the rate of 17 g. of water-vapour per square decimeter of leaf surface per hour. This is several times greater thanthe maximum rate of transpiration ever recorded for a sunflower plant. Evidently some other factor than the diffusive capacity of the stomates has a limiting effect upon transpiration whenever the stomates are widely open. The foregoing discussion of the diffusion of gases through small openings has been based on the assumption that diffusion is occurring into a quiet atmosphere. If an air current is blowing across the surface of a multiperforate septum with a sufficient velocity to prevent the formation of diffusion shells the result is a steeper diffusion gradient and hence a greater rate of diffusion of gas through the pores of the septum. Hence, when subjected to air movements, the diffusion of water through multiperforate septa or the stomates may be even greater than the preceding discussion has indicated. Within limits increase in the velocity of the wind results in a progressive increase in the rate of diffusion through a multiperforate septum.

THE MECHANISM OF THE OPENING AND CLOSING OF THE STOMATES

The degree of stomatal opening is influenced both by changes in the turgor of the guard cells and by changes in the turgor of the epidermal cells, although the former usually play a predominant role. In general an increase in the turgor of the guard cells relative to that of the epidermal cells leads to a widening of the stomatal aperture, and vice versa. The greater this turgor difference, the wider the stomatal aperture.

The mechanism of the effect of changes in the turgor of the guard cells upon the size of the stomatal aperture varies with the structure, form, and position of the stomates. In one type of guard cell, found in many different species of plants, the cell wall is thicker on the side bordering the stomatal pore than on the side bordering the epidermal cells. With an increase in turgor the thinner walls of the guard cells are stretched more than the thicker; this causes the thicker-walled sides to assume a concave shape and results in the appearance of a gap the stomatal pore between the two guard cells. Opening of the stomates typical of the grass and sedge families appears to be due to swelling of the ends of the guard cells thus separating the abutting walls of the middle portion of the two adjacent guard cells. In the sunken stomates typical of conifers opening of the stomates seems to result largely from a change in the shape of the guard cells due to an increased turgor which is unaccompanied by any appreciable stretching of the walls. These various types of stomatal mechanisms are discussed by Copeland.

The three principal factors which influence the opening and closing of the stomates are:

1. Light,
2. The internal water relations of the leaf, and
3. Temperature.

INFLUENCE OF THE LIGHT FACTOR IN STOMATAL OPENING AND CLOSING

The most familiar of all stomatal reactions is their response to light. Unless other conditions, to be discussed later, are limiting, the stomates of most species open when exposed to light, and close upon the failure of illumination. Most commonly, therefore, the stomates are open in the daytime and closed at night, although there are many exceptions to this statement. The sensitivity of stomates to the light factor probably varies considerably according to species. Within limits the stomates appear to respond quantitatively to the amount of light absorbed. Stomatal opening apparently will occur in all wavelengths of the visible spectrum, although the influence of radiations in the red region appears to be weaker than the influence of other wavelengths of the visible spectrum.

Upon the cessation of illumination the stomates usually begin to close. Generally this is a gradual process and, according to Stiil felt the greater the quantity of light which has been absorbed during the course of a day, the longer it takes for the completion of stomatal closure.

The work of Sayre Scarth and others indicates that the mechanism whereby light brings about stomatal opening and the mechanism whereby its absence causes stomatal closure are primarily osmotic, but are conditioned by changes in the H-ion concentration of the guard cells. Illumination of the guard cells has been found to result in a decrease in their H-ion concentration; failure of illumination in an increase. Thus Scarth found the pH of the guard cells of the Wandering Jew (Zebrina pendula) to range from 5.0 or less in the dark to between 6.o and 7.4 in the light. The other cells of the leaf do not change appreciably in H-ion concentration upon the advent or failure of illumination, suggesting that the guard cells are less effectively buffered than the other leaf cells. Although several explanations have been offered to account for the effects of light and darkness on the pH of the guard cells none of them are strongly supported by experimental evidence and they will not be discussed.

The guard cells apparently always contain starch, but the quantity present is not constant from one hour of the day to the next. Sayre has shown that the starch content of the guard cells is at its maximum during the night, decreases rapidly during the daylight hours, and increases again towards evening. The sugar content of the guard cells bears a reciprocal relationship to the starch content; when the latter is high, the sugar content is low, and vice versa. These reciprocal changes are apparently the result of reversible reactions in which the total amount of carbohydrate involved doesn't vary greatly. Conversion of starch to sugar and of sugar to starch results from the action of the complex of enzymes known as diastase. Increase in the pH of the guard cells such as occurs upon the incidence of light, appears to favour the hydrolytic (starch to sugar) action of diastase. On the contrary a decrease in their pH, such as occurs in the evening, favours the synthetic action of this enzyme whereby sugar is converted into starch.

Increase in the sugar concentration of the guard cells results in an increase in their osmotic pressure while a decrease in their sugar concentration has the opposite effect. That such changes in the osmotic pressure of the guard cells actually occur has been shown by many investigations. The diurnal changes in the osmotic pressures at incipient plasmolysis of the guard cells and epidermal cells.

In general the osmotic pressure of the guard cells is usually relatively high during the daylight hours, and relatively low at night. The osmotic pressure of the epidermal cells does not change appreciably during the course of the day and approximates that of the guard cells at night. Increase in the osmotic pressure of the guard cells in the morning results in an increase in their diffusion pressure deficit relative to that of the contiguous cells. Water therefore moves into the guard cells, increasing their turgor, which in turn leads to a widening of the stomata1 aperture. Similarly a decrease in the osmotic pressure of the guard cells results in a diminution of their turgor and a narrowing of the stomata1 aperture. However, certain facts suggest that the effects of light upon stomates cannot be interpreted entirely in terms of the osmotic mechanism just described. One of these is the rapidity with which stomatal opening occurs.

In some species at least the stomates open within less than a minute after exposure to light. It is difficult to visualize such a rapid action in terms of an enzymatic reaction, since usually such reactions occur at a relatively slow rate. Some investigators consider therefore that light exerts a direct effect upon the guard cells in addition to its indirect influence upon the magnitude of their osmotic pressure. Little if anything is known, however, concerning the mechanism of any such reaction.

INFLUENCE OF THE WATER FACTOR IN STOMATAL OPENING AND CLOSING

As subsequent discussion will show development of an internal water deficit in plants is of frequent occurrence, especially on warm summer days. A shrinkage in the total volume of water in a plant results in general in a diminution in the volume of water in each individual cell, although all cells will not necessarily be affected equally. Such a decrease in the water content of the leaf cells, not sufficient to induce visible wilting, is often called incipient wilting. Under such conditions the guard cells usually decrease in turgor as a result of osmotic movement of water into contiguous cells. Reduction in the turgor pressure of the guard cells as a result of the diminution of the volume of water in them will bring about a partial to complete closure of the stomates. There is also some evidence that diminution in the water content of the guard cells induces a decrease in the pH of their cell sap and a correlated conversion of sugar to starch. The resulting diminution in the osmotic pressure of the guard cells may lead to further loss of water from the into adjacent epidermal cells.

THE TEMPERATURE FACTOR IN THE OPENING AND CLOSING OF STOMATES

The results of various workers upon the influence of temperature on stomatal behaviour are not in agreement and divergent views have been expressed uponthe subject. According to Scarth relatively high temperatures (35–40°C.) accentuate opening of the stomates of Zebrina pendula in the light, and prevent closure or even induce opening in the dark. Relatively low temperatures (0–8°C.) prevent opening of the stomates of this species, even in the light. There is some evidence that the pH of the cell sap of the guard cells increases with rise in temperature and that this change in H-ion concentration is accompanied by hydrolysis of starch to sugar. Presumably the reverse transformation occurs at certain lower temperatures.

DAILY PERIODICITY OF STOMATAL OPENING AND CLOSING

The stomates of all species of plants which have been studied exhibit a more or less regular daily periodicity of opening and closing, although the behaviour of the stomates upon plants of any one species varies, often markedly, depending upon the pattern of the daily cycle of environmental factors. In the course of this book we shall have occasion to analyze the hour-to-hour variations of a number of physiological processes occurring in plants.Even in a given species the daily periodicity as these hour-to-hour variations are termed of any physiological process will vary greatly according to environmental conditions. It will therefore be convenient to choose a definite type of diurnal cycle of environmental factors as a reference standard in terms of which to discuss daily variations in the rate of various processes. For this purpose we shall select a representative summers day as our criterion. We shall consider this hypothetical day to be characterized by a clear sky, a soil adequately well supplied with water and a maximum temperature of 3Q–35O C. (86–95 F.). Furthermore we will assume that the daily march of the environmental factors ofsolar radiation, air and soil temperatures, wind velocity, and atmospheric humidity will be representative of a day upon which the conditions as defined above prevail. In our subsequent discussion we shall refer to these as standard day conditions. Such environmental conditions will actually be approximated on many summer days in temperate zone regions. It is important to realise that all of the stomates on a plant do not open at the same time. Neither do all of them close at the same time. Under most conditions, however, stomates probably are more nearly coincident in the time of their opening than in the time of their closing. The aggregate diffusive capacity of all of the stomates on a plant must therefore be thought of in terms of the two factors of the degree of opening of the individual stomates and the number of stomates that are open.

It is probable that the stomata1 mechanism of every species of plant exhibits certain distinctive features in the way in which it reacts to various combinations of environmental factors; hence only broad generalizations can be formulated regarding the daily periodicity of stomata1 behaviour. The results of a number of investigators indicate that as a rule, under the conditions prevailing on a standard-day, the stomates of most mesic species of plants are open all or most of the daylight period and closed at night their maximum diffusive capacity being attained about midday, or little before. The stomates open in the morning under the influence of the light factor and soon attain nearly their maximum diffusive capacity. Under standard day conditions, however, the water content and turgor of the leaf cells usually decrease progressively during most of the daylight period. Because of this internal water deficit which develops in the leaf, stomata1 closure usually begins about noon or a little before. Virtually complete closure of the stomates often takes place considerably before the advent of darkness because of the predominant effect of the water factor over the light factor during the afternoon hours..

Innumerable other types of daily cycles of stomatal behaviour are possible,a few of which will be described briefly. The diffusive capacity of the stomatessometimes rises to a mid or late morning maximum, decreases markedly during the midday hours, rises to a secondary maximum during the afternoon, and finally falls to a virtually zero value at approximately the termination of the daylight period. The stomates apparently behave in this way when a water deficit develops in the leaves somewhat earlier in the day than under standard day conditions. Partial closure of the stomates results during the morning hours. The resultant reduction in their diffusive capacity permits an increase in the water content of the leaf, and for a time the stomatal apertures again widen. Subsequently the water deficit of the leaf increases again, due to increased transpirational loss, and the stomates enter upon a second cycle of closing which continues throughout the remainder of the day light period. When the soil water supply is distinctly inadequate the stomates usually open incompletely and seldom remain open for the entire daylight period. Although with the advent of daylight the light factor favours stomatal opening, especially on clear days, the water content of the leaf is so low that opening is seldom complete. Furthermore the effect of the water factor usually begins to predominate over the light factor relatively early in the day and stomatal closure may be complete by mid-day or even sooner. During prolonged droughts the stomates generally close progressively earlier and earlier each day and ultimately matinal opening may cease almost entirely.On cloudy or rainy days, especially if the temperature is relatively low the stomates of most species open less completely than on clear days when the soil is well supplied with water. This is due principally to the ineffectiveness of the light factor in inducing stomatal opening under such conditions..Hence opening of the stomates is incomplete and they do not remain open as long under such meteorological conditions as on a standard day. Nocturnal opening of the stomates has been reported for a number of species of plants but the conditions which result in this type of stomatal behaviour are not clearly understood. Apparently it may be brought about by different combinations of environmental conditions and it seems likely that the conditions leading to nocturnal opening of the stomates may differ according to species. In the Norway spruce for example, the stomates remain open during the night after a cloudy day on which relatively high temperatures have prevailed. There are some indications that high temperatures and a deficient water supply favour nocturnal opening of the stomates in some species. Scarth et al. have shown that a reduced partial pressure of oxygen in the atmosphere leads to stomatal opening in the dark. They suggest that nocturnal opening may be due to a reduced partial pressure of oxygen in the intercellular spaces resulting from night respiration. Finally it should be noted that in some species such as maize it has never been possible to demonstrate night opening of the stomates.

FACTORS AFFECTING TRANSPIRATION

The rate of transpiration of a plant or any leaf on a plant varies from day to day, from hour to hour, and, frequently, from minute to minute. Variations in the rapidity with which water is lost by plants are due to the effects of environmental factors upon physiological conditions within the plant. The important environmental factors influencing the rate of transpiration are:

1. Solar radiation,
2. Humidity,
3. Temperature,
4. Wind,
5. Soil conditions influencing the availability of water, and
6. Atmospheric pressure.

This last factor is relatively much less important than the other five listed.While the general effect of variations in the intensity or magnitude of each of these factors upon transpiration is well known, and has frequently been demonstrated by experimentation, the precise mechanism of the effect of each is not so easily amenable to experimental treatment. The following interpretation of the mechanism of the effects of these environmental factors upon the rate of transpiration is therefore a somewhat theoretical one, but is in accord with the experimental data available at the present time.

SOLAR RADIATION

This term refers to the visible light and other forms of radiant energy (infrared and ultraviolet radiations) reaching the earth from the sun. The principal effects of solar radiation upon transpiration result from the influence of light upon the opening and closing of the stomates. In most of the species of plants which have been studied the stomates are usually closed in the absence of light, thus causing a virtually complete cessation of stomatal transpiration during the hours of darkness. Since none of the other environmental factors can have any influence upon stomatal transpiration except when the stomates are open, light occupies a position of prime importance among the environmental conditions influencing transpiration.

A second important effect of solar radiation upon transpiration operates through its influence on leaf temperatures. In direct sunlight the temperature of leaves is almost invariably higher than the air temperature. This effect will be analyzed later in this chapter.

HUMIDITY

Several units are used for designating the humidity conditions of an atmosphere. One of these is the actual vapour pressure of the atmosphere. This should not be confused with the saturation vapour pressure. Since the rates of diffusion and evapouration are influenced directly by the vapour pressure of the atmosphere this is usually the most satisfactory unit in which to express humidity values for physiological purposes.

The vapour pressure deficit is also a common, and for many purposes a valuable unit for the expression of humidity values. The vapour pressure deficit is the difference between the actual vapour pressure of the atmosphere and the vapour pressure of a saturated atmosphere at the same temperature. As an example, consider an atmosphere having a vapour pressure of 10.51 mm. Hg at 250 C. The saturation vapour pressure of this atmosphere would be 23.76 mm. Hg. Hence its vapour pressure deficit is 13.25 mm. Hg (23.76 10.51). If an evapourating surface and the atmosphere are both at the same temperature the vapour pressure deficit indicates directly the steepness of the vapour pressure gradient between the atmosphere and the evapourating surface. The term vapour pressure gradient as applied to water-vapour has the same significance as the term diffusion pressure gradient which has been frequently used in previous discussions. In fact the vapour pressure gradient of water vapour is its diffusion pressure gradient. As long as the temperature of the evapourating surface does not differ materially from that of an atmosphere the rate of evapouration from a saturated surface into that atmosphere will show a close proportionality to its vapour pressure deficit. However, the temperature of an evapourating surface is seldom exactly the same as that of the atmosphere and the deviation in temperature between the two is often very considerable. There are therefore many situations in which rates of evapouration do not show a close proportionality with the vapour pressure deficit of the surrounding atmosphere.

The most generally employed of all units in expressing humidity values is the relative humidity, which is the percentage saturation of an atmosphere at a given temperature. For example, a saturated atmosphere at 30 C. will have a vapour pressure of 31.82 mm. Hg. The relative humidity of this atmosphere would be 100 per cent. If only half the amount of vapour is present that would be present at saturation at this temperature (i.e. a vapour pressure of 15.91 mm. Hg) then the relative humidity would be 50 per cent etc. The relative humidity depends both on the concentration of water-vapour in the atmosphere and upon the temperature. Any increase in temperature with no accompanying change in the amount of water-vapour present always results in a decrease in relative humidity, because an increase in temperature means that a higher vapour pressure will be required before the atmosphere is saturated.

Expression of humidity values in terms of relative humidity, although a common practice, is unsatisfactory for physiological purposes because the same relative humidity, 50 per cent for example, may refer to widely different vapour pressures and vapour pressure deficits. A relative humidity of 50 per cent at 10 C. is equivalent to a vapour pressure deficit of 4.60 mm. Hg, while a relative humidity of 50 per cent at

500 C. is equivalent to a vapour pressure deficit of 46.26 mm. Hg. In other words,with the same relative humidity, evapouration from moist surfaces exposed tothe air will be many times as rapid at 5Q0 C. as at 100 C. Only when all of the relative humidity values are recorded at the same temperature are they an expression of the relative differences in the vapour pressure of an atmosphere.

In general the greater the vapour pressure of an atmosphere, other factors remaining unchanged, the slower the rate of transpiration. The rate of diffusion of water-vapour out of a leaf depends upon the difference between the vapour pressure in the intercellular spaces and the vapour pressure of the outside atmosphere, since the vapour pressure is a measure of the diffusion pressure of the water-vapour. Let us suppose that the vapour pressure of the intercellular spaces is 31.82 mm. Hg which is the value for a saturated atmosphere at 30 C. Such a vapour pressure is frequently attained in the internal air spaces of leaves. Let us further assume that at the same time the vapour pressure of the atmosphere is only half as great (15.91 mm. Hg). In very quiet air the vapour pressure in the neighbourhood of transpiring leaf surfaces may be greater than that in the atmosphere in general but in this discussion it is assumed that there is sufficient air movement to prevent any appreciable accumulation of water-vapour in the vicinity of the leaves. Under the conditions as stated diffusion of water-vapour would occur through the stomates at a relatively rapid rate. If the vapour pressure of the atmosphere were lowered below this value the rate of diffusion of water-vapour out of the leaf would be increased; conversely increase in the vapour pressure of the atmosphere would result in a decrease in the diffusion rate of water-vapour out of the leaf. Similarly, an increase in the vapour pressure of the intercellular spaces would result in an increase in the rate of transpiration while a decrease in the vapour pressure of thc intercellular spaces relative to that of the atmosphere would have the converse effect. On the rare occasions when the vapour pressures of the atmosphere and of the intercellular spaces are equal, no transpiration will occur, even if the stomates are open. The final statement in the preceding paragraph, should not, however,be interpreted to mean that transpiration can never occur into a saturated atmosphere. If the temperature of the leaf is higher than that of the surrounding atmosphere this will, if the intercellular spaces be saturated with water-vapour, result in~ a higher vapour pressure in them than in the atmosphere, even when the latter is also saturated with water-vapour. Under such conditions outward diffusion of water-vapour occurs, resulting in a localized region of supersaturated atmosphere around the leaf in which condensation of water-vapour may take place.

TEMPERATURE EFFECTS ON TRANSPIRATION

THERMAL RELATIONS OF LEAVES

While leaf temperatures often do not deviate greatly from surrounding atmospheric temperatures the discrepancy between the two is frequently sufficiently great to make it necessary to take it into account in careful experimental work. Leaves exposed to direct solar or artificial radiation usually have temperatures of from 2 to 100 C. (sometimes even more) in excess of that of the atmosphere. Under other conditions, to be described later, leaf temperatures may be less than that of the atmosphere.

Leaf temperatures are generally measured by means of thermocouples. A thermocouple is made by twisting together the ends of two fine wires of dissimilar.

Theoretically the temperature of a leaf may be regarded as conditioned by four different influences:

1. Thermal absorption,
2. Thermal emission,
3. Internal endothermic (energy-storing) processes, such as photosynthesis and transpiration, and
4. Internal exothermic (energy-releasing) processes such as respiration.

The influence of this last factor upon leaf temperatures is practically always negligible and will be disregarded. Similarly the quantity of energy transformed in photosynthesis is relatively so small that it need not be considered in evaluating the factors determining the temperature of leaves. Thermal emission refers to the loss of heat from a leaf by the processes of conduction, convection, and radiation. Thermal absorption refers to the gain of energy by a leaf by these same physical processes. Heat transmission which is brought about by intermolecular contacts is known as (thermal) conduction. The greater the temperature difference between a leaf and its environment the more rapidly conduction of heat will occur from the leaf to the gases of the atmosphere, if its temperature is the higher, or in the opposite direction if the temperature of the atmosphere is the higher.

Whenever loss of heat is occurring by conduction from a leaf to adjacent gas molecules of the atmosphere convection currents are set up in the atmosphere in the vicinity of the leaf. Cooler gas will displace the gas in the vicinity of the leaf surfaces which has become warmed as a result of thermal conduction from the leaf. This accelerates the rate at which conduction can occur from the leaf.

Radiation is the transfer of radiant energy across space. The best known types of radiant energy are light, infrared radiations (heat waves) and ultraviolet radiations. Radiant energy is commonly pictured as being propagated across space in the form of undulatory waves. Radiation occurs from the molecules of one body to those of another only if the radiating body is at a higher temperature than the receiving body. Thus light infrared and ultraviolet radiation are transferred from the sun to the earth metals, for example copper and constant an (an alloy of copper and nickel). A difference of electrical potential is set up between two wires thus brought into intimate contact. The magnitude of this electrical potential is very nearly directly proportional to the temperature of the thermocouple. In actual practice two junctions are generally used, one being inserted into the leaf blade, the other being kept at a standard temperature, usually 0^0C. The difference in potential between the two thermocouples is generally measured with a potentiometer.

TEMPERATURE EFFECTS ON TRANSPIRATION

In like manner a warm stove loses heat by invisible infrared radiation to its environment. Radiation occurring from leaves is also in the infrared range of wave lengths.

The temperature of a leaf exposed to direct sunlight or strong artificial illumination almost invariably exceeds that of the atmosphere. Thermal absorption in this case direct absorption of solar radiations proceeds under such conditions at a rapid rate. A portion of the absorbed radiant energy is dissipated (usually) by transpiration, and a portion is lost from the leaf by thermal emission. As already shown transpiration alone is never adequate to dispose of the energy absorbed if the leaves are exposed to strong insolation. Thermal emission under these conditions will involve loss of heat both by radiation and by conduction. In general the greater the excess of leaf temperature over that of the atmosphere the larger the proportion of heat loss by thermal emission as contrasted with transpiration.

On a cloudy day the temperature of leaves seldom deviates very greatly from that of the enveloping atmosphere. Heat exchanges between a leaf and its environment under such conditions probably occur principally by conduction. Similarly at night leaf temperatures usually do not deviate greatly from those of the surrounding atmosphere.

Leaves sometimes have a lower temperature than the surrounding atmosphere. This is often true, for example, of leaves which are transpiring fairly rapidly but which are not exposed to direct sunlight. Such leaves are often cooler by several degrees centigrade than the atmosphere. It is also possible that leaves sometimes lose heat energy by direct radiation to their environment rapidly enough to result in a lowering of their temperature below that of the atmosphere. This is especially likely to occur on clear nights when the vapour pressure of the atmosphere is low. These conditions favour direct radiation from the leaves to the sky, i.e. to relatively cold gases especially carbon dioxide and water-vapour.

The temperature of leaves usually fluctuates from minute to minute especially during daylight hours. Minor fluctuations are due largely to shifts in wind velocity. The thermal emissivity of leaves exposed to direct sunlight is invariably increased by an increase in wind velocity. When the sunlight is intermittent, frequent changes in leaf temperatures also occur. Each time the sun is obscured by a cloud a sudden drop in leaf temperature usually takes place. Contrariwise, when the sun emerges from behind a cloud there is usually a distinct and abrupt rise in leaf temperature. The factors controlling the internal temperatures of the other organs of plants are in general similar to those influencing leaf temperatures. The temperatures of fleshy leaves, fruits, tree trunks, succulent stems such as those of cacti, etc. under the influence of direct insolation may often greatly exceed those of the surrounding atmosphere. The side of an apple fruit exposed to direct sunlight, for example, may have a temperature of from 12 to 25 C. higher than the air temperature.

THE INFLUENCE OF TEMPERATURE UPON TRANSPIRATION RATES

The effectsof temperature upon the rate of stomatal transpiration can be most clearly analyzed in terms of its effect upon the difference in vapour pressures between the intercellular spaces and the outside atmosphere. In this part of the discussion it is again assumed that the air movement is sufficient to prevent any appreciable accumulation of water-vapour in the vicinity of the leaf surfaces. Suppose that the temperature of both the leaf and the surrounding atmosphere increases from 20 C. to 300 C. Unless the leaf is markedly deficient in water this will result in an increase in the vapour pressure of the intercellular spaces from approximately 17.54 mm. Hg to approximately 31.82 mm. Hg, these being the values for a saturated atmosphere at 200 C. and 300 C., respectively. The atmosphere of the leaf intercellular spaces is in direct contact with the relatively extensive evapourating surface of the mesophyll cell walls, hence the vapour pressure in the intercellular spaces tends to remain in equilibrium with the water in the mesophyll cells. In order to simplify this part of the discussion, it will be assumed that the atmosphere of the leaf intercellular spaces maintains essentially a saturation vapour pressure for the prevailing leaf temperature. Under certain conditions, as shown later in this chapter, maintenance of even an approximately saturated atmosphere throughout the intercellular spaces probably does not occur, but for the time being this possibility will be disregarded.

In the surrounding atmosphere, however, vapour pressure conditions are very different. On clear days, that is, on the very type of day upon which the highest rates of transpiration occur, there is frequently little change in the vapour pressure of the atmosphere over land surfaces during the course of a single day 2 Evapouration into the atmosphere is insufficient to permit a rapid building up of the vapour pressure towards the value for a saturated atmosphere as the temperature of the air increases during the day. This statement should not be misinterpreted to read that the vapour pressure of the atmosphere is invariable. The magnitude of the vapour pressure of the atmosphere varies greatly from day to day and from season to season, depending upon the prevailing climatic conditions. On cloudy or rainy days, the atmospheric vapour pressure is generally greater than on clear days during the same season; in the summer months it is generally greater than in the winter months, etc. Nevertheless the statement made above that on clear, bright days there is often little change in the vapour pressure of the atmosphere is essentially correct that the process of transpiration itself, occurring on a grand scale from a vegetation-covered area of the earths surface would be sufficient to increase the vapour pressure of the lower layers of the atmosphere during the daylight hours. The atmosphere is so vast, however, in relation to the amount of water-vapour lost by plants, that over short periods, transpiration has only a slight effect on its vapour pressure except in deep ravines or other local habitats in which movement of the air is restricted.

Increase in the temperature of the atmosphere does result in an increase in the speed of the water-vapour molecules present. If the volume of the atmosphere remained constant this would result in a small increase in its vapour pressure. But even this effect is never fully realised in the atmosphere because an increase in temperature also results in an expansion of the atmosphere entirely or largely offsetting its influence in increasing vapour pressure. If we assume for the purpose of our specific example that the vapour pressure of the atmosphere at 200 C. was half that of a saturated atmosphere at that temperature S.77 mm. Hgthen the excess vapour pressure of the leaf over that of the atmosphere at 30 C. was 8.77 mm. Hg (17.54 8.77). At 300 C., however, the vapour pressure of the intercellular spaces would have increased to about 31.82 mm. Hg while the increase in the vapour pressure of surrounding atmosphere would usually be so slight that it can be disregarded. The excess vapour pressure of the intercellular spaces over the atmosphere is now 23.05 mm. Hg (31.82 8.77) which will result in diffusion of water-vapour out of the leaf at a rate nearly three times as fast as at 200 C. The effect of a rise in temperature therefore is principally an increase in the steepness of the diffusion gradient (vapour pressure gradient) of water-vapour through the stomates, and hence an increase in the rate of transpiration.

So far we have considered only examples in which the temperature of a leaf and the surrounding atmospheres are the same. We have already noted however, that the temperature of leaves exposed to direct sunlight is almost always higher than that of the atmosphere. If the temperature of a leaf is increased above that of the surrounding atmosphere by the absorption of solar radiation, the usual effect is an increase in the magnitude of the excess vapour pressure of the intercellular spaces over that of the outside atmosphere. Let us continue the example which has been presented earlier. At 30 C. under the conditions as stated the vapour pressure difference between the intercellular spaces and the atmosphere was about 23.05 mm. Hg. Suppose, however, that due to the absorption of radiant energy, the temperature of the leaf is 350 C.while that of the atmosphere remains at 300 C. Water would evaporate from the walls of the mesophyll cells until the vapour pressure in the inter cellular spaces approximates that of a saturated atmosphere at 350 C. (42.18 mm. Hg). The gradient between the intercellular spaces and the atmosphere is therefore increased to about 33.41 mm. Hg (42.18 8.77) and the rate of transpiration is correspondingly increased. The sudden increases in transpiration rate which are often observed when plants are shifted from shade to direct sunlight, as for example in potometer experiments, are undoubtedly due largely if not entirely to effects of this type. Similarly it has been observed that a sudden decrease may occur in the transpiration rate when a passing cloud obscures the sun, which is followed by just as sudden an increase when the sun again emerges from behind the cloud. For further analysis of the effects of leaf temperature upon transpiration rates.

WIND

The effect of wind upon the rate of transpiration is far from simple, and depends in part upon the other prevailing environmental conditions. Usually, however, increase in wind velocity, within limits, results in an increase in the rate of transpiration. This is usually explained by assuming that water-vapour often accumulates in the vicinity of transpiring leaves in a quiet atmosphere, especially if they are not exposed to direct sunlight. The result of such an accumulation of water-vapour is a decrease in the steepness of the vapour pressure gradient through the stomates and hence a decrease in the rate of transpiration. If, however, the leaves are exposed to a wind, any accumulation of water-vapour molecules in the immediate vicinity of the leaf surfaces will be dispersed. The effective result will be an increase in the steepness of the vapour pressure gradient through the stomates, and a consequent increase in the rate of loss of water-vapour. Whenever a temperature differential exists between a leaf and the surrounding atmosphere, convection currents are set up in the gases in the vicinity of a leaf which may largely or entirely prevent any accumulation of water-vapour in the immediate vicinity of the leaf. The influence of wind in raising the transpiration rate of leaves is probably more effective, therefore, when they are subjected to such conditions that their temperature does not depart appreciably from that of the surrounding atmosphere.

The effect of wind in dispersing accumulated water-vapour in the vicinity of the leaf surfaces is probably of less importance in its influence upon transpiration rates than its effect in causing the swaying of branches and shoots and the bending, twisting, and fluttering of leaf blades. It has been shown experimentally that immobile leaves usually transpire less than similar leaves allowed to bend and move freely when both are exposed to wind of equal velocity. Such bending and contortion of leaves may increase the rate of water-vapour loss in part by compressing the intercellular spaces, thus forcing water-vapour and other gases out through the stomates.

Chapter 13

Availability of Soil Water

A gentle breeze is relatively much more effective in increasing the transpiration rate than winds of greater velocity. Winds of very high velocity have been observed to have a retarding effect upon transpiration. This is probably due to closure of stomates under such conditions. The effect of wind upon the transpiration of a leaf which is exposed to intense insolation is further complicated by the fact that a wind may have a cooling effect upon the leaves, due to their increased thermal emissivity under such conditions. Reduction in the temperature of a leaf, as we have already seen, exerts a decelerating effect upon transpiration. If conditions are such that this cooling effect of wind is predominant increased wind velocity may result in a diminution in transpiration rate.

SOIL CONDITIONS INFLUENCING THE AVAILABILITY OF WATER

Although transpiration can continue for short periods at rates considerably in excess of the rate of absorption of water, in general, if soil conditions are such that absorption of water is appreciably retarded, the rate of transpiration will soon show a corresponding retardation. The availability of soil water to the plant is therefore an important and, in fact, often the limiting factor in transpiration. The principal soil factors which affect the rate of absorption of water by plants are:

1. Available soil water,
2. Soil temperature,

3. Aeration of the soil, and
4. Concentration of solutes in the soil solution.

All of these factors indirectly influence the rate of transpiration.

FACTORS AFFECTING TRANSPIRATION

ATMOSPHERIC PRESSURE

It has been demonstrated experimentally that a reduction in atmospheric pressure results in an increase in the rate of transpiration. This result would be predicted theoretically since reduction in the density of the atmosphere would be expected to permit diffusion of water-vapour to occur into it more rapidly. In any given locality variations in atmospheric pressure are too slight to have any significant effect upon the rates of transpiration. Plants growing at high altitudes are subjected to distinctly lower atmospheric pressures thanthe plants of lowlands, and in comparative evaluations of the transpiration rates of species growing in these two types of habitats the influence of this factor must be considered.

STRUCTURAL FEATURES OF PLANTS WHICH INFLUENCE THE RATE OF TRANSPIRATION

A number of diverse species of plants growing under identical environmental conditions usually differ very greatly in transpiration rates. Physiological factors, such as differences in stomatal behaviour, cell sap concentration, colloid content of the cells, etc. are undoubtedly partially responsible for the fact that all plants do not lose water-vapour at the same rate even under the same environmental conditions. Structural differences, particularly in the leaves, also account in part for unlike rates of transpiration of different species growing in the same environment. Many of the supposed effectsof structural differences in plants upon their transpiration rates are, however based solely on inferences, and not upon experimentally determined facts. Many species which on the basis of their anatomy have been judged to have a low transpiration rate later were actually found to transpire very rapidly when environmental conditions were suitable. The following discussion of the anatomical features known or generally believed to influence the rate of transpiration will therefore be very brief.

THICKNESS OF THE CUTIN.

It can be easily demonstrated that water vapourates more readily from uncutinized epidermal cell walls than from those which are coated with a layer of cutin. It is sometimes assumed, therefore that the thicker the layer of cutin on the outer walls of the epidermis of a leaf, the slower the rate of cuticular transpiration from that leaf. Actually,however, it seems obvious that any increased thickness in the cutin layer beyond that at which cuticular transpiration becomes negligible will have no appreciable influence on the rate of evapouration from the epidermal cell walls.

EPIDERINAL HAIRS.

H airs are prominent features of one or both of the surfaces of the leaves of many species. Living hairs probably increase the rate of cuticular transpiration by increasing the exposed surface of the leaf. Many botanists have assumed that dead hairs reduce the rate of stomata1 transpiration, particularly under conditions of intense sunlight or strong winds. In the former case it has been supposed that the whitish hairs reflect such a large proportion of the incident light that the leaf temperature is not as high as it otherwise would be, and that the rate of transpiration is correspondingly lower. In the latter case it has been supposed that the hairs impose a mechanical barrier to the effect of wind on transpiration. However, Sayre found that removal of the hairs from the leaves of the mullein had little effect upon the rate of stomatal transpiration under ordinary conditions of light intensity and wind velocity, although such treatment did result in an increase in cuticular transpiration.

RATIO OF INTERNAL TO EXTERNAL SURFACE IN LEAVES

Although transpiration rates are most frequently expressed in terms of the exposed area of the leaf surface, most of the water-vapour lost from leaves evapourates from the walls of the mesophyll cells which bound the intercellular spaces. The area of this internal evapourating surface in proportion to the external surface of a leaf varies greatly not only in leaves of one species as compared to those of another, but in leaves of the same species if they have developed under different environmental conditions. Computations of the ratio of internal exposed surface to external exposed surface of leaves of a number of species have yielded values ranging from 6.8 to 31.3. In general the greater the proportion of palisade to sponge tissue the greater this ratio.The thickness of the leaf will also obviously be a factor influencing this ratio.The ratio is greater in sun leaves than in shade leaves of the same species; for the lilac the values are 13.2 and 6.8, respectively.

It has generally been supposed that the transpiration rates of leaves in which a relatively large area of mesophyll cell walls is exposed to the intercellular spaces as compared with the exposed epidermal area of the leaf are greater than in leaves in which the contrary condition obtains. Theoretically such an effect could be exerted only if the area of exposed mesophyll cell walls exercised a controlling influence upon the steepness of the vapour pressure gradient through the stomates. When transpiration is relatively high a steeper vapour pressure gradient may be maintained by leaves possessing relatively extensive internal evapourating surfaces than by those which do not. The higher transpiration rate of sun leaves as compared with shade leaves on the same plant, which has been observed in some species, may be partly explainable on this basis.

SUNKEN STOMATES.

The stomates of many species of plants are sunken below the general level of the epidermis. Since the length of a diffusion gradient is one of the factors governing its steepness, diffusion through a small opening is slower if the gas must pass through a relatively long tube before reaching the orifice, than if the diffusion is through a shorter tube. In leaves in which the stomates are not sunken the diffusion column is relatively short, amounting only to the diameter of the guard cells as seen in cross section, while in sunken stomates the column of gas may be longer by several times. The diffusion of gases through such stomates is undoubtedly slower than through stomates of the ordinary type. This retarding effect of sunken stomates upon the rate of transpiration as compared with the stomates of the non-sunken type has been estimated by Renner to range from 30 to 70 per cent.

STRUCTURE AND DISTRIBUTION OF THE ROOT SYSTEM.

Under some environmental conditions an individual plant of a given species will develop a greater leaf surface in proportion to the extent of its root system thanunder others. The latter type of structural development will, other conditions being equal, favour the maintenance of higher transpiration rates thanthe former.

The distribution of a root system in the soil sometimes influences rates of water absorption and transpiration. In a habitat where deep-rooted and shallow-rooted species are growing side by side the former may transpire more rapidly during drought periods than the latter because their roots penetrate to a soil horizon which still contains available water, while roots of the latter are in a moisture deficient soil. The completeness with which the soil mass is interpenetrated by roots may also influence rates of absorption and transpiration. Thus Miller found that sorghum plants have almost twice as many fibrous roots as maize plants. Rates of absorption and transpiration can be better maintained, especially when the soil is relatively dry, by plants with the sorghum type of root system than by plants with the maize type of root system.

THE DAILY PERIODICITY OF TRANSPIRATION

All plants exhibit a daily periodicity of transpiration rate which varies somewhat with the species and is greatly influenced by the environmental conditions to which the plant is exposed. We shall first consider the transpiration periodicity upon a standard day as defined in the preceding chapter, since most experiments upon daily variations in transpiration rate have been conducted upon plants growing under approximately such conditions. It should not be assumed, however, that under natural conditions the daily periodicity of transpiration usually or normally follows the trend. The daily periodicity of transpiration under these roughly defined conditions does, however, afford a convenient reference standard with which to compare the many variations which are manifested by this phenomenon.

The transpiration rate during the hours of darkness is generally low, and in most species water loss during this period may be regarded as almost entirely cuticular. It is not justifiable to assume that absolutely no stomatal transpiration occurs at night in any species unless this is definitely known to be true. In some species of plants, as we have already seen, nightstomatal opening is a common occurrence, while in others it occurs under certain environmental conditions. Even in those species in which the stomates are normally closed during the hours of darkness, a few may remain open. The transpiration rate shows a rapid rise during the early morning hours followed by a slower but consistent increase which culminates in a point of maximum transpiration which occurs sometime during the mid-day hours. Following this peak of transpiration, the rate decreases usually consistently although often with an accelerated rate in the late after noon, until the low and virtually steady night rate is attained at approximately the termination of the daylight period.

The daily periodicity in the rate of stomatal transpiration in most species of plants can be interpreted almost entirely in terms of two variable factors:

1. The diffusive capacity of the stomates, and
2. The vapour pressure gradient between the intercellular spaces of the leaf and the outside atmosphere. In analyzing the daily periodicity of transpiration it is the diffusive capacity of the stomates in the aggregate (i.e. all of the stomates on a plant) which must be considered. The aggregate diffusive capacity of the stomates on a plant on which all of the stomates were half open, for example,

would be much greater than that for a similar plant on which half of the stomates were fully open and half fully closed.

With the advent of daylight the stomates open over a period of time which varies in length according to the species and with environmental conditions. Some open earlier or more rapidly than others. The initial rise in the rate of transpiration in the morning is due to the opening of the stomates, resulting in a gradual increase in their aggregate diffusive capacity. As each stomate opens a vapour pressure gradient is established through it between the atmosphere of the intercellular spaces and the outside atmosphere. During the hours of darkness the leaf cells regain most or all of the water which they have lost during the preceding day, and thus attain approximately their normal turgidity. This facilitates the evapouration of water into the intercellular spaces, so that when the stomates open in the morning be internal atmosphere of the leaf is usually saturated with water-vapour, or practically so. The vapour pressure of the intercellular spaces is therefore at the maximum possible for that leaf temperature. This is, on clear days, almost always in excess of the vapour pressure of the atmosphere when the stomates open, and often considerably so. Hence outward diffusion of water-vapour through the stomates usually begins as soon as they are open.

Once a stomate is well open, minor variations in the size of the stomatal aperture have little or no effect on the rate of water-vapour loss through it.The rate of transpiration continues to increase, however, for some time afterthe stomates, in the aggregate, have attained their maximum diffusive capacity. This is due to a gradual increase in the steepness of the vapour pressure gradient through the stomates. As the day progresses the temperature of both the atmosphere and the leaf increases; if the latter is in direct sunlight its temperature is invariably somewhat in excess of that of the atmosphere. On clear days, as previously described, the vapour pressure of the intercellular spaces usually increases relative to that of the atmosphere with a rise in temperature. The steepness of the vapour pressure gradient between the internal atmosphere of the leaf and the external atmosphere therefore usually increases progressively during the earlier part of the day, and this is the important factor accounting for a rise in the transpiration rate once the stomates have attained approximately their maximum diffusive capacity.

Almost from the moment when the stomates begin to open in the morning a train of events is set in operation in the plant which ultimately causes a reduction in the rate of transpiration. During approximately the first half of the day, however, these factors are more than offset by the factors resulting in an increase in the rate of transpiration. In most plants, while transpiration is occurring rapidly, the rate of absorption of water does not keep pace with the rate at which water-vapour is lost from the leaves. This results in a reduction in the water content of the entire plant, and especially that of the leaves. In more extreme cases wilting results, but under standard day conditions the leaves seldom pass beyond the stage of incipient wilting which corresponds only to a partial loss of turgor by the leaf cells.

THE DAILY PERIODICITY OF TRANSPIRATION

Decrease in the water content of the leaf cells also results in a rise in their diffusion pressure deficit, and a correlated decrease in their vapour pressure. A fully turgid cell (i.e. one with a diffusion pressure deficit of zero) has an equilibrium vapour pressure equivalent to that of pure water regardless of its osmotic pressure. At 20 C. this is 17.54 mm. Hg. If the diffusion pressure deficit of the mesophyll cells of a leaf at this temperature were raised to 100 atmos., a value which can be attained in only a very few species of plants, their vapour pressure would decrease by only about 1.27 mm. Hg. Since the vapour pressure of the cell walls, with which the vapour pressure of the intercellular spaces tends to be in equilibrium, in turn tends to be in equilibrium with that of the cell sap, it would appear that an increase in the diffusion pressure deficit of the mesophyll cells has only a very slight influence upon vapour pressure conditions in the intercellular spaces.

However, certain other factors must be taken into consideration in evaluating the effect of a diminished leaf water content upon the vapour pressure of the intercellular spaces. Decrease in the water content of the leaf cells involves a reduction in the quantity of water in the cell walls as well as in other parts of the cell. Water undoubtedly passes through cell walls principally through the intermicellar material. The principal effect of a decrease in the quantity of water in the walls is a shrinkage in the volume of the hydrophilic intermicellar material and a resulting contraction in diameter of the intermicellar capillaries. This probably results in a decrease in the permeability of the walls to water. Hence, under such conditions, water often may not pass across the walls with sufficient rapidity that an equilibrium vapour pressure is maintained between the cell sap and the outer surface of the cell walls. When the water content of a leaf is low the vapour pressure of the cell wall surface in contact with the intercellular spaces may therefore be less than the vapour pressure of the water in the cell sap. Any such reduction in the vapour pressure of the intercellular spaces results in decreasing the steepness of the vapour pressure gradient between the intercellular spaces and the outside atmosphere, and hence decreases the rate of transpiration.

The experimental results of That indicate that the vapour pressure of the intercellular spaces is often at less than a saturation value. During the course of a day, however, the steepness of the vapour pressure gradient through the stomates may be decreased in still another and perhaps more important way. When the stomates open in the morning the vapour pressure gradient between the saturated internal leaf atmosphere and the outside atmosphere is short and hence steep, being approximately equal in length to the depth of the guard cells. For some time after the stomates open essentially such a condition is probably maintained. As the day advances the rapid loss of water-vapour molecules out of the intercellular spaces, perhaps coupled with a gradual reduction in the vapour pressure of the cell walls makes it less and less likely that a condition of saturation can be maintained throughout the intercellular spaces. The zone in which a vapour pressure inequilibrium with the water in the mesophyll cell walls is maintained almostcertainly retreats more and more deeply into the intercellular spaces. Eventually it may be restricted to a thin layer just above the evapourating surfaces of the mesophyll cell walls. This gradual lengthening of the vapour pressuregradient through the stomates is probably an important factor in bringing about a reduction in the rate of transpiration during the afternoon hours. It has frequently been observed in studies of transpiration periodicity that the transpiration rate often begins to decrease during the midday period before any appreciable change occurs in the area of the stomatal apertures.This indicates that the initial reduction in transpiration rate is induced by some other factor than a change in the diffusive capacity of the stomates. As the preceding discussion has shown this other factor is almost certainly a reduction in the steepness of the vapour pressure gradient through the stomates. By late afternoon the air temperature and the intensity of the solar radiation begin to decrease appreciably, thus inducing a decrease in the temperature of the leaf. This lowering of the leaf temperature may further decrease the vapour pressure of the intercellular spaces, and hence further depress the steepness of the vapour pressure gradient, since temperature changes have very little influence on the vapour pressure of the outside atmosphere under standard day conditions. Thus the effect of a reduction in the vapour pressuregradient through the stomates in decreasing the rate of transpiration, which first becomes apparent during the mid-day period, continues with augmented effect as the hours of darkness approach.

Continued diminution in the leaf water content due to the excess of transpiration over absorption eventually results also in a diminution of the turgor of the guard cells due to a decrease in the water content of the leaf. This results in a gradual closure of the stomates. Some of the stomates on a plant probably begin to close even before the time at which the peak of the transpiration rate is attained. In all likelihood stomates near the margin or tip of a leaf begin to close before those in the middle since the effects of deficiency of water usually appear first in these regions of a leaf. With an increasing leaf water deficit more and more of the stomates on the plant close. This soon results in a reduction in the aggregated iffusive capacity of the stomatal population of the plant. As the afternoon advances more and more of the stomates become completely closed, resulting in a progressive diminution in diffusive capacity. This gradual closing of the stomates, resulting in an increasingly larger proportion of them being closed as the day advances, is the principal factor reducing the rate of transpiration during the latter part of the day. The effect of this factor and that of a decreased steepness of the vapour pressure gradient overlap during a large part of the afternoon.

Finally, by late afternoon, complete closure of essentially all of the stomates has occurred. Stomata1 transpiration is terminated at that time and during the hours of darkness the rate of water loss from the plant is controlled by factors influencing the rate of cuticular transpiration. Under environmental conditions deviating very greatly from those which were postulated in the preceding discussion, transpiration periodicity curves may be entirely different from those. Variations in temperature, light intensity, humidity, and soil water supply may all markedly influence both the trend of transpiration periodicity, and the magnitude of the daily water loss.

Low temperatures may result in a complete elimination of stomatal transpiration by inducing stomatal closure. Low light intensities, such as those existing on cloudy days, are unfavourable to stomata1 opening in most species. The stomates seldom open completely under such conditions and the period during which they are open is usually of shorter duration than on clear days. Furthermore, the vapour pressuregradient through the stomates is seldom as steep on such days as on clear bright days, since leaf temperatures never appreciably exceed atmospheric temperatures except when exposed to direct sunlight, and atmospheric vapour pressures are usually higher during cloudy days than on clear days at the same season of the year. The magnitude of transpiration under such conditions is usually greatly reduced, and transpiration periodicity curves plotted for plants exposed to such conditions usually present a somewhat fore shortened and greatly flattened appearance.

A deficient soil water supply is most commonly the factor which causes marked departures from the type of transpiration periodicity already considered, especially during the summer months. The effect of a gradual diminution in soil water content upon transpiration periodicity. A reduction in soil water content has two pronounced effects upon the daily march of transpiration. The total daily magnitude of water loss is decreased, and the peak of the transpiration curve often occurs somewhat earlier in the day than under conditions of abundant soil water supply. Since, even in temperate regions, periods of decreased soil water supply are of common occurrence during the summer months, and in many habitats are the rule rather than the exception for many species of plants much of the time. Internal factors may also be responsible for transpiration periodicities of a different trend from those which have already been discussed. In some species of plants, as has already been described, the stomates remain open to a greater or lesser extent during the hours of darkness. Such plants have higher transpiration rates at night than those in which the stomates are closed. In some species of cacti there is a complete inversion of the usual transpiration periodicity curve, transpiration rates being regularly greater at night than in the daytime. This appears to be due to complete or nearly complete stomatal closure during the hours of daylight, while the stomates are, as a rule, open at night.

SEASONAL VARIATIONS IN TRANSPIRATION RATES

In temperate regions transpiration occurs predominantly during the warmer months of the year and especially during those periods of the warm season when the soil water supply is abundant. Among the deciduous group of woody perennials, the branches are defoliated during most of the autumn, all of the winter, and the earlier part of the spring. Although the twigs and branches which remain exposed to the atmosphere are completely encased in corky layers of bark and the buds are enclosed within cutinized bud scales some water loss occurs from such species even during the winter months. Young twigs lose water under such conditions faster than older ones. Winter transpiration rates of deciduous woody plants are always negligible in comparison with summer rates. In evergreen species of either the needle-leaved or broad-leaved type the transpiration rates are usually not appreciably different at most times during the winter from the transpiration rate of deciduous trees at that season.

This is probably due principally to the fact that the low temperatures of the winter prevent opening of the stomates although the relatively low leaf water contents characteristic of this season may also be a factor in maintaining the stomates in a closed condition. Mild periods of any considerable duration often have a detrimental effect upon evergreen species during the winter months. The warm air temperatures induce stomatal opening, and a relatively high rate of transpiration ensues. The soil is apt to be deficient in water at such times, or even if present, that in the surface layers may remain frozen during a period of warm air temperatures. Even if neither of the foregoing conditions exists, the soil temperatures will be low, and this greatly retards the rate of absorption of water.

The combined effect of a relatively high transpiration rate and relatively low absorption rate results in a gradual desiccation of the leaves and branches of the plant. This diminution in the water content of the aerial organs during warm periods in winter is more severe in windy weather and is more likely to occur if a sequence of mild days follows immediately after a cold spell. If severe enough this desiccation will result in the death of some of the branches, or in extreme cases of the entire tree or shrub. This is one cause of the phenomenon known as winter-killing. Winter-killing due to desiccation of the tissues should not be confused with cold injury due to low temperatures which is an entirely different phenomenon.

THE MOVEMENT OF WATER THROUGH THE PLANT

Most terrestrial plants obtain the water which is necessary for their existence from the soil. An overwhelmingly large proportion of the water which is absorbed by the roots of land plants is lost in the process of transpiration. Smaller quantities are utilized in growth and in photosynthesis and in some species limited amounts of water may be lost by guttation. Water must therefore move through the intervening tissues and organs from the absorbing regions of the root to the tissues in which it is utilized, or from which it passes out of the plant. The process whereby water moves through the plant is often termed the conduction, transport, or translocation of water.

In herbaceous species and many shrubby plants the distance throughwhich water moves in passing from the root tips to the leaves is usually not more than a few feet. Even in such plants appearances are sometimes deceptive, as some herbaceous and shrubby species such as alfalfa may have such deep root systems that some of the absorbed water often ascends for distances as great as twenty or more feet before it reaches the level of the soil surface. It is in trees, however, in which the most striking illustrations of the upward movement of water occur. The tallest tree of which we have an authentic record is a specimen of the Coast Redwood which has attained a height of 364 feet. Many other individuals of this species and of several others, including the Big Trees of California, the Douglas Firs (Pseudotsuga mucronata) of the Pacific Northwest, and the Blue Gums (Eucalyptus) of Australia exceed 300 feet in height. Trees ranging from 100 to 200 feet in height were of common occurrence in the virgin forests of eastern North America. Since the root systems of trees always penetrate at least a few feet into the ground the actual vertical distance through which at least a part of the water absorbed is elevated is always more than the height of the tree. In trees, therefore, water must ascend to heights ranging up to nearly 400 feet above the level of water absorption.

The mechanism by which this feat is accomplished in tall trees has been the subject of much experimentation, and even more speculation. The ensuing discussion of the ascent of sap through plants will be presented largely in terms of its movement through trees, principally because much of the experimental work on this problem has been performed on woody species. Any explanation of this phenomenon which can be shown to be adequate for tall trees should also prove satisfactory for vascular species of lesser stature.

THE PATH OF WATER THROUGH THE PLANT

Water enters the plant through the epidermal cells and root hairs at or near the tips of the roots and crosses the cortex, endodermis, and a part of the pericycle before it finally enters the lumina of the vessels or tracheids of the root xylem. At this point its upward movement begins. The xylem tissue is continuous from just back of the root tips, through the roots, into and through the stems, the petioles of leaves, and ultimately, usually only after much branching, terminates in the mesophyll of the leaf. The xylem tissue through which the water moves is thus a continuous unit system within the body of the plant. Along most of its course water moves en masse through the vessels or tracheids of the xylem. At the termination of the xylem ducts in the leaves the water passes into the adjacent mesophyll cells. In the mesophyll of the leaves it moves from cell to cell, eventually most of it being lost from the cells by ognized at least since the time of the girdling experiments of Malpighi in 1671. Girdling (or ringing) a stem, so that all of the tissues external tothe xylem are removed, does not prevent movement of water to organs attached to that stem above the ring. On the contrary cutting through the xylem tissue of a stem results in almost immediate wilting of leaves attached to the stem above the ring.

ANATOMY OF STEMS

Since the mechanism of the movement of water can scarcely be understood intelligently without some knowledge of the anatomy of the tissues through which it moves, a brief review of stem structure is desirable before proceeding further with a discussion of this process. Stems vary greatly in their structure, every species possessing some anatomical features which are peculiar to itself. Nevertheless certain general patterns of tissue arrangement have been found to prevail, and the stem structure of most species will approximate one or another of these general arrangements. The corn stem represents an herbaceous monocot type of structure while that of the tulip poplar is typical of woody dicot stems. The structure of woody stems cannot properly be appreciated merely from a consideration of one year old stems. The primary tissues of such stems develop from tissues formed at the apical growing tip during growth of the stem in length. Nearly all perennial stems also grow in diameter as a result of the development of secondary tissues from the cambium. By successive stages of division, enlargement, and differentiation of cambium cells additional layers of xylem (secondary xylem) are laid down on the inner face of the cambium, and new layers of phloem tissue (secondary plzloem) on its outer face Hence after a few years the great bulk of any woody stem is composed of secondary tissues.

Secondary xylem and secondary phloem also develop from the cambium in most herbaceous dicot stems but in such species the cambium in any one of phloem and xylem elements from the cambium. (A) a cambium initial, (B) division of cambium cell, (C) outer cell resulting from division becomes a phloem element, (D) another division of the cambium cell, (E) outer cell resulting from second division becomes a xylem element. Actually several xylem or several phloem elements are often formed successively. Upon maturation most of the xylem and phloem elements acquire sizes and shapes which are very different from those of the cambium initials from which they originate. Stem is never active for more than one growing season. The spring formed xylem tissue, as viewed in cross section is usually distinctly different in aspect from that formed later in the season. In many angiosperms the springwood contains more and larger vessels and the cell walls are generally thinner than in the subsequently produced summer wood. In the conifers the spring formed tracheids are thinner-walled and of larger cross-sectional diameter than those formed later in the growing season. The transition from spring to summer wood is often a very gradual one. On the other hand the more open xylem tissues formed each spring abut directly upon the denser tissues which were produced during the preceding summer, thus giving rise to an abrupt line of demarcation between the zones of xylem formed in any two successive seasons. The result of this growth behaviour is that a cross section of the trunk or a branch of any tree appears as a system of concentric layers, the so-called annual rings, each representing an annual increment of growth. In rare cases no annual rings nor more than one annual ring may be produced in a season, but usually each ring represents the xylem produced by the activity of the cambium during one season.In many woody species as the xylem tissues increase in age important changes occur in the colour, composition, and structure of the various elements, resulting in the conversion of sapwood into heartwood. As sapwood ripens into heartwood the walls of any remaining living cells of the xylem become increasingly lignified death of these cells soon following. The water content of the tissues is generally reduced and such compounds as oils, resins gums, and tannins accumulate in the cells or cell walls. The darker colouration of the heartwood of most species as compared to the sapwood is due to such accumulations. In mature trees the heartwood becomes merely a central supporting column surrounded by a cylinder of sapwood which varies in thickness from a few to many annual layers, depending upon the species and the environmental conditions under which the tree is growing. In some species (apple, elm) the heartwood remains virtually saturated Tangential section of a small portion of the wood of whitepine (Pinus strobus). The vertically oriented elongated elements are tracheids. Bordered pits show in sectional view in the tracheid and marginal ray cells. Simple pits show in sectional view in other ray cells. Half-bordered pits show between marginal ray cells and other ray cells. The water in the heartwood of such species as apple and elm appears to be largely static and is not directly involved in translocation. Coincident with the development of secondary xylem, secondary phloem tissues also develop from the cambium. Cork cambiums are also initiated in the bark which produce cork layers. Profound modifications therefore occur in the outer tissues as well as in the xylem of woody stems as they grow older.

A somewhat more detailed concept of the structure of the cells and elements of the stems of angiosperms through which the movement of water occurs is presented which represent longitudinal tangential and radial sections from the xylem of a tulip tree which may be taken as representative for this group of plants. The xylem tissues of the wood of angiosperms are composed of vessels tracheids, fibers of several types, wood parenchyma, and xylem ray cells. There is a tremendous variability in the proportional distribution and arrangement of these tissues according to species.

The most characteristic elements of the xylem tissue of angiosperms are the vessels. These are, in general, more or less tubular structures which may extend through many feet of the xylem. In some species cross walls, usually perforated, are of frequent occurrence in vessels, in others such cross walls are infrequent or lacking. In diameter they may range in trees from about 20 to about 400. In vines they may be as much as 700 in diameter. The vessels branch extensively in certain regions of the plant, especially at nodes within the leaf lamina, and at the points in the root systems where rool branching occurs.

The vessels of the proto xylem (the first cells of the xylem to mature during the ontogeny of a growing stem or root tip) have cellulose walls which are reinforced by distinctive lignified thickenings which appear as rings, spirals,or other characteristic patterns. The vessels that develop later in the ontogeny of any growing tip have lignified, usually pitted walls which lack any of the thickenings which distinguish the walls of the vessels of the protoxylem. Pits normally occur in all parts of vessel walls which are contiguous with other vessels or cells. These pits consist essentially of thin areas in the walls. The walls of the vast majority of plant cells are pitted. The architecturally distinct type known as the bordered pit.

ANATOMY OF STEM

Tracizeids are found in wood of many but not all species of angiosperms. They are typically more or less spindle-shaped cells with thick walls which are almost always lignified. As viewed in cross section they are usually angular. Mature tracheids contain no protoplasm and hence are non-living like the xylem vessels. The largest tracheids are about 5 mm. in length, and about 30 in diameter, although most of them are smaller. The walls of tracheids are pitted like those of vessels. Tracheids are water-conducting cells but in most angiosperms they are relatively of much less importance as channels throughwhich water moves than the vessels.

Tracheids are developed from the cambium by a process essentially similar to that by which vessel segments are formed, except that the size and shape of the resulting cells is different, and that there is no coalescence of the individual elements such as occurs in the formation of vessels. The xylem of the angiosperms also contains fibers of various types, which in general are similar to the tracheids in structure except that usually they have thicker walls, smaller lumina, and fewer and smaller pits. Fiber cells are non-living and their walls are lignified. Because of their thick walls and small lumina the fibers do not play an important part in the movement of water through stems.

The xylem tissue of practically all angiosperms also contains wood parenchyma cells which, unlike the elements already described, remain alive for some time after their differentiation. Usually death of the wood parenchyma cells does not occur until the wood of which they are a part is converted into heartwood. Wood parenchyma cells are generally somewhat elongate, and roughly four-angled in cross section. They occur in the xylem as vertical series of cells placed end to end. The strands of wood parenchyma thus produced often extend for a long distance in a vertical direction through the wood. The distribution of the wood parenchyma strands throughout an annual ring of xylem tissue varies with the species. In some they are more or less scattered through the xylem, in others they occur only in the last layer or two of cells which are produced in the summer woodin other words at the termination of the seasons growth and in others only in contact with the vessels, or in contact with other wood parenchyma cells which are themselves in contact with the vessels.

All of the xylem elements previously described are oriented with their long axes in a vertical direction. In addition to this vertical system there is also present in the xylem a transverse, radiating system in which the long axis of the cells is at right angles to the long axis of the stem. These transversely oriented tissue units are known as vascular rays. In the stems of most species they are continuous from the outer extremity of the phloem through the cambium into the xylem, which they penetrate to a greater or lesser distance. The portion of the vascular ray found in the xylem is termed the xylem ray; that found in the phloem the pliloem ray. Xylem rays may vary from one to many cells in thickness and likewise in height, certain types usually being characteristic of any one species. The cells of the xylem rays,like those of the wood parenchyma, usually remain alive until the woody tissue is converted into heartwood. The cells of the xylem ray are typically elongate and more or less angular in cross section. The xylem rays probably serve as routes along which lateral movement of water occurs from the xylem to the cambium and phloem, and along which translocation of soluble foods takes place from the phloem to the living cells of the xylem.

The living ray cells are in contact at various points with the strands of living wood parenchyma cells. The vertically oriented wood parenchyma strands and transversely oriented xylem ray strands thus form a unit system of living cells within the woody cylinder. Hence there is present a continuous inter meshing network of living cells throughout the greater mass of non-living vessels, tracheids, and fibers in the younger portions of any woody angiosperm stem. There are probably few if any of the conducting elements vessels and tracheids which are not in contact at one or more points with this continuous system of living cells.

The wood of the gymnosperms is simpler in its structure than that of the angiosperms, this group of woody species also showing, in general, a greater uniformity in stem structure than the latter group. The only cell types universally present in the wood of coniferous trees are the tracheids and woodray cells. Wood parenchyma cells are also present in the wood of most species of conifers, while in many species tracheid-like fiber cells are also present. The most important distinction between the wood of conifers and that of angiosperms is the total absence of vessels in the former.The tracheids are the distinctive element in conifer wood, constituting as they do the great bulk of all the woody tissues present in such species. In conifers the tracheids form a densely packed type of woody tissue, composed of interlocking cells. Vertically contiguous tracheids always overlap along their tapering portions. Movement of water and solutes from one tracheid to another is facilitated by means of the bordered pits in the adjacent walls. Because of the numerous cross walls in a xylem tissue composed almost entirely of tracheids, water encounters a greater resistance in moving through such tissues than in traversing woody tissues which contain vessels. Nevertheless it is interesting to note that the tallest trees in the world are conifers in which all movement of water occurs through tracheids.

Theories of the Mechanism of the Movement of Water through Plants. A number of different theories of the mechanism by which the ascent of water is brought about iii plants have been suggested, and it is probable that more than one mechanism is involved in this process. The present state of our knowledge justifies a discussion of only three possible mechanisms:

1. That ascent of sap is caused by the vital activities of living cells, principally those in the stem,
2. That this process occurs according to a mechanism described in the cohesion of water theory, and
3. That upward movement of water occurs as a result of root pressures. In passing, it should be noted that neither atmospheric pressure nor capillarity can account for the rise of sap in plants, although both of these processes are recurringly suggested as being involved in this process. Atmospheric pressure could never account fora rise of more than about 10 meters, and most mature trees attain a greater height than this. The capillary rise of water in tubes 10 in diameter is only about 3 meters, and few if any of the conducting elements have as small a diameter as this.

The mechanism of the movement of water through the plant is inextricably bound up with the mechanism of the absorption of water, so the following discussion will necessarily treat to some extent of both of these processes. Vital Theories. While the vessels and tracheids through which longitudinal transit of water occurs are non-living, they are always in more or less intimate contact with living cells. Suggestions have therefore been made from time to time that the motive power causing movement of water is furnished by the living cells of the stem. The most recent advocate of such a theory is Bose.

As an example of the vital theories of the ascent of sap we may consider the proposals advanced by Godlewski in 1884. He considered that conduction of water occurred as a result of periodic changes in the osmotic pressure of the living cells present in the xylem (especially the wooJ ray cells). Upon increase in their osmotic pressure water was supposed to move into the cells from bordering xylem elements. Upon decrease in osmotic pressure water was supposed to move back into a xylem duct. In the xylem duct it was supposed that water would move upward until it came in contact with another living ray cell because of a lower air pressure in the upper than in the lower part of the vessels. The ascent of water in stems was thus supposed to be brought about by a number of repetitions of this same process. There is little or no direct evidence in support of this or any of a number of more or less similar theories which have been suggested.

The experiments of Strasburger demonstrated quite clearly that the primary mechanism of the rise of sap in trees operates independently of the living cells of the stem. He performed many experiments upon the movement of water through woody stems in which the living cells had been killed by one method or another. In one experiment, for example, he used a 75 -year -old oak tree about 22 meters in height. This was sawed off close tothe ground and the cut end transferred to a solution of picric acid, which is toxic to living cells. The picric acid solution slowly moved up the stem. Fuchsin, added to the liquid in which the base of the tree was immersed 3 days after the picric acid, also ascended to the top of the tree through tissues in which the living cells had been killed by the picric acid. Water also continued to ascend through similar stems after they had been completely killed by exposure to a temperature of 900 C. Similar experiments have been performed by later investigators. All of these investigators have confirmed Strasburgers results that water would continue to ascend for some time through stems, segments of which had been exposed to one treatment or another which would kill all living cells present. In all experiments of the type just described, it has been observed that the leaves at the top of a stem, part or all of which has been killed, sooner or later wilt and wither, although this effect is by no means an immediate one,often appearing only after several days. Proponents of the vital theories have accepted this as evidence that the living cells of a stem are essential for the passage of water through it. Dixon however, considers that this delayed lethal effect on the leaves is due to either or both of two causes. Killing of the stem tissues often causes the formation of substances which plug up the vessels or tracheids, thus impeding the upward movement of water. Furthermore, death of the cells in the treated regions causes the release into the conducting channels of toxic compounds which, when transported to the eaves, cause death of the leaf cells. There is still some doubt, however,whether all of the retardation in rate of translocation as a result of such experiments can be explained in these two ways. The possibility remains open that the living cells of the xylem may in some way be necessary to the maintenance of the water-conducting capacity of the xylem, or even that they may contribute in some way to the lifting of water through stems.

In general, the evidence appears to be quite conclusive that the living cells of the stem do not furnish the principal mechanism which causes the upward transport of water in stems. On the contrary, as shown in the subsequent discussion, the living leaves or stem tips at the top of the plant appear to be absolutely necessary for the ascent of water through stems. Root Pressure. That a slow exudation of sap often occurs from the cut surface of a stem is well known. Such sap exudations are often ascribed to the development of a pressure in the dilute sap of the xylem vessels resulting from the operation of some as yet not fully understood mechanism in the root cells. Hence the name root pressure. It is generally considered that such pressures may occur in intact plants as well as in those which have been cut into.

It is probable that not all of the phenomena which have been classed under the name of root pressure are due to the operation of the same physiological mechanism. Some sap exudations seem to come from the cambium or phloem rather than from the xylem. It seems quite certain that some of the sap exudations from the xylem occur from the living wood parenchyma or woodray cells rather than from the vessels, or at least that the exudation is greatly influenced by the activity of such cells. Hence the living cells of the stemas well as those of the root are probably involved in many root pressure phenomena. James and Baker go so far as to regard the subjection of the sap in vessels to a pressure as an exceptional occurrence in plants, but this seems to be an extreme point of view, and will not be followed in this discussion. Because of the unsettled state of our knowledge regarding this phenomenon it will be necessary to use the term root pressure in a very uncritical sense.

The magnitude of root pressures is commonly determined by attaching a manometer to the cut surface of the plant from which the bleeding is occurring. With only a few rare exceptions the exudation pressures which have been recorded for plants do not exceed 2 atmos., and most of them are less than this. According to White individual excised roots of tomato plants exude sap from their basal ends with a pressure of more than 6 atmos. It has not yet been shown, however, that such pressures operate for any considerable distance through plants. In general there is no correlation between the pressure with which the sap is exuded from cut stems and the volume of sap flow. In some species relatively large volumes of sap may be exuded under a relatively low pressure, in others exactly the opposite relation may exist.

Some of the woody plants in which bleeding from cut surfaces occurs,commonly ascribed in part or entirely to root pressure, are the sugar maple,box elder, dogwood, hornbeam (Ostrya), sycamore, birches, currant, and grape. Such exudation of sap under pressure usually occurs only during the seasons of the year when the plant bears no foliage, and most commonly in the early spring months. The popular concept of a rise of sap in woody plants in the spring is largely based on the observation of such phenomena. Exudation of sap under pressure can also be demonstrated in many species of herbaceous plants especially if potted plants are used and the soil is flooded with water before the demonstration is attempted. The sap which moves through woody plants in the spring and, less abundantly, in the autumn under the influence of root pressures contains appreciable concentrations of carbohydrates as well as traces of mineral salts. In contrast with this is the condition of the xylem sap in the summer at which season it contains traces of mineral salts, but practically no soluble carbohydrates.

It has sometimes been claimed that root pressures are adequate to account for the rise of water in herbaceous plants and in low shrubs or trees,and that they may even play a very considerable part in the ascent of sap in taller trees. While it is undoubtedly true that root pressure does, in some species of plants, under certain conditions, account for the movement of some water in an upward direction through plants, this process is inadequate to account for the translocation of any considerable amount of water through plants. In the first place there are many species in which this phenomenon has never been observed. In the second place, the magnitude of the pressure developed is seldom sufficient to force water to the top of any except relatively low-growing species of plants. Neither is the rate of flow, as it occurs inmost species, anywhere near rapid enough to compensate for the known rates of transpiration. Finally, and this is perhaps the most fundamental objection to the idea that root pressures play a very prominent part in the ascent of sap in plants; they are usually negligible, in temperate regions at least,during the summer period when transpiration is most rapid. During periods of rapid transpiration, the cut surfaces of plants not only fail to exude sap, but will usually absorb water if it is supplied at the cut surface.

The trunks of many species of deciduous trees exhibit a marked seasonal variation in water content. A minimum water content is attained during middle or late summer and the trunks gradually fill up during the fall and spring. Often there is a temporary shrinkage in their water content during the winter months but the main period of diminishing trunk water content occurs in the spring and summer. Some authorities even speak of a seasonal rise and fall in the water table of a tree trunk, and consider that intercellular spaces as well as cells become occupied with water during this process. Root pressure undoubtedly accounts at least in part for such seasonal replenishments of the store of water in tree trunks.

The living cells of the xylem probably play an active part in this process. It is possible that there may be slowly ascending streams of water in the living cells of the xylem which operate entirely independently of the water columns in the vessels or tracheids. Hence the same portions of the xylem which at times contain gases may at other seasons be occupied by water. The trunks of many conifers show no marked seasonal variations in water content; this is possibly correlated with the lack of appreciable root pressures in such species. The Cohesion of Water Theory. The leading advocate of this theory has been Dixon who has performed much of the experimental work upon which it is based. A number of other workers, notably Askenasy and Renner have also contributed important theoretical and experimental evidence in support of this theory. Molecules of water, although ceaselessly in motion, are also attracted to each other by strong cohesive forces. In masses of liquid water the existence of such cohesive forces is not obvious, but when water is confined in long tubes of small diameter the existence of attractive forces between water molecules can often be demonstrated. If the water at the top of such a tube be subjected to a pull the resulting stress will be transmitted all along the column of water, due to the mutual attraction between the molecules.

The water conducting system of plants constitutes just such a system, enclosing continuous thread-like columns of water which extend from the top to the bottom of the plant. A stress applied at any point to this system will be propagated to all its parts. The application of such a stress stretches the water into taut threads, throwing it into a condition of tension, which is the equivalent of negative pressure. According to this theory forces develop in the upper parts of plants, and especially in the leaves, which cause the rise of water through the plant. As evapouration proceeds from the walls of the mesophyll cells into the intercellular spaces the diffusion pressure deficit of the mesophyll cell walls increases. Water therefore moves into the walls from the adjacent protoplasm,this resulting in turn in the movement of water from the cell sap into the protoplasmic layer. The resulting increase in the diffusion pressure deficitof the cell sap is in turn propagated to the protoplasm and the cell sap in all parts of the cell. The osmotic pressures of the leaf cells are generally high enough that diffusion pressure deficits of sufficient magnitude to account for movement of water to the top of the plant can develop in them without a complete loss of turgor by the leaf cells. If the cell is in direct contact with one of the branches of the xylem ducts which ramify throughout the lamina of the leaf, water will move from the vessel or tracheid into the cell this results in the development of tension in the water column terminating in this xylem element. The diffusion pressure deficit of the water in the conducting elements will be increased by the amount of this tension. Water under a tension (negative pressure) of 10 atmos., for example, has a diffusion pressure just 10 atmos. less than that of pure water at the same temperature which is not under tension. In other words its diffusion pressure deficit is 10 atmos. If the cell from which evapouration is occurring is not in direct contact with a xylem element a gradient of diffusion pressure deficits, gradually increasing in magnitude from cell to cell in the direction in which the water is moving, is established between the xylem element and the cell from which evapouration is occurring.The tension developed in the conducting elements is transmitted along their entire length to their lower termination just back of the root tips and probably, very often at least, across the root tissues as well. The magnitude of the tension which develops in the xylem conduits is increased by conditions which favour a rate of water loss from the leaves considerably in excess of the rate at which water enters the roots from the soil.

The movement of water across the cells of the root from the soil into the lower ends of the water-conducting elements must be considered as an integral part of the translocational process. At their lower terminations thexylem vessels or tracheids are in contact with the pericycle, or more rarely,directly with the endodermal cells of the root. Water moves from the adjacent root cells into the conducting elements because the tension (diffusion pressure deficit) developed in these elements exceeds the diffusion pressure deficit of the cells of the root. When the tension in the water columns is relatively low a gradient of diffusion pressure deficits will be established across the root cells similar to those which are established in the leaf mesophyll cells when water is moving through them. The diffusion pressure deficits will increase from cell to cell along this gradient in the direction in which water is moving, i.e. from the periphery of the root towards the xylem. Since, however, the osmotic pressures of root cells are usually lower than those of the mesophyll cells it is probable that the tension developed in the water columns often exceeds the highest osmotic pressures of any of the root cells through which water passes on its way from the soil to the xylem. Under such conditions the volume of water in the root cells may continue to diminish even after their turgor has been reduced to zero. As a result the walls of the cell are pulled inwards due to the adhesion between them and the contracting mass of water.

The counter pull exerted by the elastic walls of the cell will stretch the encompassed mass of water and throw it into a state of tension. Whenever a tension of sufficient magnitude has been generated in the conducting elements, therefore,the water in the root cells may also pass into a state of tension. Even under such conditions a gradient of diffusion pressure deficits would be established across the root, but the mechanism responsible for the development of this gradient would be more complicated than were no tension generated in the cells. When the water in a cell is under tension its diffusion pressure deficit is equal to its osmotic pressure plus the tension to which the water is subjected.

Regardless of the exact mechanism involved, the essential fact is that the development of tensions in the water columns induces the establishment of a gradient of diffusion pressure deficits, increasing consistently from cell to cell across the root from its peripheral cell layer to the conducting elements and water will move along this gradient from the epidermal layer to thexylem tissue. Water enters the peripheral cells of a root whenever the diffusion pressure deficit of water in the soil is less than that of the water in the epidermal cells and root hairs in the absorbing zone of the root.

In most trees practically all upward movement of water occurs in the vessels or tracheids of a few of the outermost annual rings of the sapwood and in some species, especially those with ring-porous wood is almost entirely confined to the outermost ring. As is well known, in some species of trees (beech, sycamore, etc.) the heartwood of the trunk may disappear completely by decay. The fact that such hollow trees continue to live and thrive is conclusive evidence that the heartwood plays no essential role in the upward movement of water in such species.

The water-conductive system of plants should be regarded as a unit system; the individual threads or columns of water not operating separately but each as a part of a more or less complicated meshwork. The air-plugged tracheids or vessel sections become merely islands in the meshwork of intact water columns, and the capacity of the tissues as a whole to conduct water, although somewhat impaired by the presence of air in some of the conducting elements, is by no means destroyed.

MAGNITUDE OF THE COHESIVE FORCE OF WATER

If water is to move through the xylem ducts as postulated by this theory, its cohesive force must be of sufficient magnitude to resist the stresses which are imposed upon it.The maximum height to which water ever moves in plants does not exceed 400 feet. This height is equivalent to about 2 atmos. which represents the maximum stress which the water columns could develop as a result of their own weight. Water, however, encounters a certain amount of resistance in moving through the conducting tissues. The greater the velocity of the current of water, the greater the resistance which it will encounter. Dixon has shown that at the velocity corresponding to usual rates of transpiration this resistance is approximately equal to the pressure required to support water for the height of the plant. He used the wood of the yew for his determinations, in which higher values for resistance would be expected than in non-conifers in which most water traverses vessels rather than tracheids. For high rates of conduction Huber estimates the resistance to be several times as great as this. Accepting Dixons estimates as an average value,the required cohesive force of water must be doubled, giving a value of 24 atmos. To this must be added the resistance normally encountered by water in crossing the tissues of the root and the mesophyll cells of the leaves which is equivalent to only a few atmospheres. The estimated minimum cohesive force required to lift water to the tops of the tallest tree is therefore about 30 atmos. Even if we assume a resistance three times as great as that found by Dixon, the value of the necessary cohesive force becomes only about 50 atmos.

Values experimentally determined for the cohesive force of water range up to 350 atmos. Dixon obtained values as high as 207 atmos. for sap centrifuged from branches of holly (hexaquifoliurn) The cohesive force of water is therefore much in excess of the minimum required for the lifting of water to the tops of even the tallest trees. Although the estimates given in the preceding paragraph represent the minimum cohesive force actually required for the movement of water to the top of tall trees, many botanists believe that when an internal water deficit exists in plants, tensions as great as 100 atmos. or even more may develop in the water columns of at least some species of plants.

One of the most ingenious methods of studying the cohesion of water employs the sporangia of the ferns for this purpose. Around the edge of a fern sporangium occurs a ring of dead, thick walled cells, known as the annulus. As the sporangium matures water is gradually lost from the cells of the annulus by evapouration. Because of the powerful adhesion between the cell walls and water, the resulting shrinkage in the volume of water in each of these cells results in the thin outer wall of each cell being drawn inward, while the ends of the thicker lateral walls are pulled toward each other. This results in tearing open the weak side ofthe sporangium, and eventually, in a complete inversion in the position of the annulus, exposing the spores on its outer surface. The annulus is now in a condition similar to that of a highly strained spring.

THE MOVEMENT OF WATER THROUGH THE PLANT

Eventually the cohesive force of the water in the cells of the annulus is exceeded, the water ruptures, and the annulus recoils suddenly into approximately its original position, scattering the spores into the atmosphere by its violent rebound. The cohesive force of the water in cells such as these can be measured by allowing the annuli to come to equilibrium with known vapour pressures, and determining the highest vapour pressure at which rupturing of the water occurs. The osmotic pressure corresponding to this vapour pressure is equal to the cohesive force of the water in the annulus cells. Such measurements indicate the cohesive force of the water in these cells to be, at a minimum, between 300 and 350 atmospheres.

The first experiment of this type was performed by Askenasy. As evapouration proceeds from the porous clay cup of this apparatus water moves up the vertical glass tube, followed by mercury from the reservoir. If the demonstration is successful the mercury will continue to rise above the level to which it will stand in a barometer. The water is now being pulled up the tube, a phenomenon which is possible only because of its cohesive force. The water in the tube is thus thrown into a state of tension which is transmitted to the mercury column below it, due to the relatively enormous adhesive force between water and mercury. The forces exerting this pull originate at the evapourating surface of the cup, and are due to the adhesive and cohesive forces operative in the maintenance of innumerable microscopic menisci in the pores of the clay cup. That succeeded in demonstrating a rise of mercury to a height of 226.6 cm. in such an apparatus. This is approximately three times as high as a column of mercury will be supported by atmospheric pressure acting alone. While the maximum tension developed in the water column in these experiments does not exceed two atmospheres, the demonstration graphically illustrates a physical system which is analogous to that which is believed to operate in the plant.

Experiments similar to that just described have also been performed using small branches or twigs of plants as an evapourating surface instead of a porous clay surface. These are attached at the upper end of a glass tube in place of the clay cup; otherwise the set-up for such an experiment is similar to that just described. While it has not been possible to obtain as great a rise of mercury in these experiments as when a porous clay evapourating surface isused, upward traction of mercury for distances well in excess of atmospheric pressure can be demonstrated if a suitable technique is followed.

DEVELOPMENT OF TENSION IN THE WATER COLUMNS

Convincing evidence that the water in the xylem vessels is often in a state of tension has been obtained by direct observations of vessels under a microscope. The stems of some species of herbaceous plants, especially the Cucurbits, are especially suitable for such observations. Such a stem of an intact, rapidly transpiring plant can be fastened in position across the stage of a microscope and, by careful dissection, vessels examined individually. If one of the vessels under observation be jabbed with the point of a fine needle, an immediate jerking apart of the water column at the point of the rupture will be seen to occur, indicating that the water in the intact vessel was actually in a state of tension. Interesting evidence that the water in the xylem ducts of woody stemsis, at times at least, under tension has been obtained by means of an instrument known as the dendrograph. This is a self-recording instrument which measures variation in the diameters of tree trunks. It is so constructed that its sensitivity is very great and that its recordings are not influenced by temperature effects upon the instrument. Dendographs are used principally to measure periodic variations in the diameter growth of trees. However, even in trees in which diameter growth has ceased, slight diurnal periodic variations in the diameter of trees have been found to be of regular occurrence. A record of the periodic variations in the diameter of a tree trunk for aperiod of several days at a season when little diameter growth was occurring. The trunk attained its minimum diameter during the transported through plants, be they the tallest of trees or herbs only a few feet in height. It is also generally agreed that root pressures play a minorrole in this process at least under certain conditions and in some species.There is no doubt that living cells are involved in the phenomena generally described under the term root pressure and there is at least a possibility that living cells of the xylem are in some way essential to the maintenance or operation of the cohesion mechanism.

RATES OF MOVEMENT OF WATER THROUGH PLANTS

The rate which water ascends through the xylem ducts may vary from a movement so slow as to bealmost imperceptible to a speed of at least 75 cm. per minute. By rate is meant the length of the water column which will move past a given point in one minute. In general daily variations in the rate of ascent of water closely parallel daily variations in transpiration rate.

LATERAL MOVEMENT OF WATER

A cell to cell lateral movement of water in a radial direction undoubtedly occurs along the vascular rays in the stems of most species of plants. In woody stems there is probably also a lateral movement of water around the stem in a tangential direction. Except in trees in which the grain of the wood is twisted, the conducting vessels on one side of the tree generally connect with branches on that side of the treeat their upper extremity, and with roots on the same side of the tree at their lower extremity. If no lateral movement of water occurred in woody stems,it would be expected that removal of the roots from one side of a tree would result in a dearth of water, or perhaps even death of the leaves or brancheson that side of the tree. Experiments have been performed on apple, peach, oak, and other woody species in which the roots on one side of the plant were removed in an attempt to determine whether or not lateral movement of the water occurs. Although the water content and growth of the plants treated in this manner diminished there was no difference in the moisture content of the leaves on the two sides of the tree. Neither did the leaves on the side from which the roots had been removed show any greater tendency to wilt on clear warm days than those on the other side. These results indicate very strongly that lateral movement of water occurs in woody stems, and that the water conductive system of plants acts as a unit system.

DOWNWARD MOVEMENT OF WATER

There are a number of experiments on record which indicate that downward movement of water can occur in stems. For example, Dixon found that if the tip of the leaf of a potato plant be cut off under an eosin solution the liquid will descend through the xylem tissue of the leaf, petiole, and stem, and eventually reach the under ground organs. In general, however, such an effect is to be expected only when an internal water deficit exists in a plant.

The cohesion theory of the movement of water will account equally well for the conduction of water in either the upward or the downward direction through the plant. If conditions are ever such that the diffusion pressure deficit of the cells of roots or some other basal organ of a plant exceeds that of the apical parts of the same plant a reversal of the direction of movement of water in the xylem will occur. Although downward movement of water can be readily induced in plants under certain experimental conditions, such movements of water are probably of infrequent occurrence in intact plants.

SOILS AND SOIL WATER RELATIONS

Most vascular plants are rooted in the soil from which they obtain both water and mineral salts. The entrance of water from the environment into any of the organs of a plant is called the absorption, intake, or uptake of water. In most vascular species the quantity of water absorbed through organs other than the roots is of negligible importance. Any thorough consideration of the entrance of water into roots requires a consideration of the properties of soils, particularly as they affect the movement of water from the soil into the root.

CONSTITUTION OF THE SOIL

The soil matrix which is the habitat of most roots is an extremely complex system. In general five different components of this system are distinguished:1. The Mineral Matter of the Soil.The parent substance of practically all soils is rock, of which there are numerous varieties. By various weathering processes rock strata are reduced to fragments of diverse sizes and these compose the bulk of most soils. The rock particles in soils may vary in size from stones and gravel down to sub-microscopic particles of colloidal clay. The mineral portion of the soil is customarily classified into several fractions depending upon the size of the particles. Such classifications usually disregard the very large particles of the soil (rock fragments, pebbles, etc.). The proportions of the different fractions present are very different in different kinds of soils.The clay fraction of the mineral portion of soils requires special consideration. Unlike the coarser fractions which are composed of small fragments of unmodified rock minerals such as quartz, feldspar, and mica, the clay portion of soils is made up almost entirely of the products of chemical weathering of the rock minerals and hence differs not only in its physical state, but in its chemical composition from the particles in the coarser fractions. Most of the particles in the clay fraction are of colloidal dimensions and exhibit the characteristic properties of colloidal systems. The particles of the clay complex, like those of many other colloidal systems, retain water by imbibition, i.e. within the structure of the particle. On the contrary the sand and silt particles of the soil retain water only on their surfaces. The presence of a considerable proportion of clay in a soil therefore endows it with a high water retaining capacity. Changes in the water content of colloidal clay result in marked changes in its volume. One result of such changes in volume is the commonly observed cracking of a soil rich in clay upon drying. The plasticity and cohesiveness of soils are also due very largely to the colloidal clay present.

Like other colloidal systems colloidal clay is markedly sensitive to the influence of electrolytes. The micelles of colloidal clay usually bear a negative charge when in contact with water. In the presence of calcium ions the individual clay particles are more or less completely flocculated into compound particles. Much of the colloidal clay in most soils exists as enveloping films around the larger soil particles and is also closely associated with organic material in the soil. Flocculation of the clay particles therefore usually results in the formation of compound granules including sand and silt particles and organic matter in addition to the clay. These are called soil crumbs. For agricultural purposes a well developed granular structure of the soil is highly desirable since such a structure favours both a high moisture-retaining capacity and a good aeration of the soil. Calcium soils, that is, those with a high calcium content, are in general the most suitable and most valuable for agricultural purposes partly because calcium favours the development of a crumb structure. In the presence of an excess of univalent cations (Na+, etc.), on the other hand, a greater proportion of the clay fraction of the soil disperses into its ultimate particles, and the soil has a single grain structure. In this condition the soil is in the least desirable structural condition for agricultural purposes. Many soils contain hydrogen ions in excess. Such soils often develop a good crumb structure, but the crumbs are less stable than those developed in a calcium soil. The addition of lime improves the physical structure of such soils. The crumb structure of a soil, especially in its surface layers, can also be destroyed by purely mechanical effects, such as trampling,or beating by heavy rains.

The Organic Matter of the Soil. Most soils contain organic matter which has been derived principally from the partial decomposition of plant residues. Small quantities may also originate from animal residues and excretions. The proportion of organic matter present may vary from almost none as in some sand deposits to 95 per cent or better in some peat soils. In ordinary agricultural soils the amount present seldom exceeds 15 per cent. In forests organic matter comes from falling leaves, dead branches and trunks of woody plants, roots which die and decay underground, and from dead herbaceous vegetation. In grasslands the underground roots and rhizomes as well as the aerial parts of the plants all contribute their quota to the organic matter of the soil. In well managed agricultural soils attempts are made to maintain the organic matter content by supplying them with organic fertilizers.

The organic matter is the seat of most of the microbiological processes occurring in the soil. One of the most important of these is the oxidation of the organic matter, a process due largely to the metabolic activities of bacteria and fungi, although a limited amount of purely chemical decomposition probably also occurs. Under conditions which are exceptionally favourable for the activities of microorganisms the organic matter of the soil is oxidized completely and disappears. For this reason the organic matter of soils in tropical regions, particularly when under cultivation, is very low. Even in more temperate regions cultivation of the soils generally results in a rapid reduction inorganic matter content, due principally to the better aeration induced by till age.

As a result of the decay process there is present in most soils organic matter in various stages of decomposition. A large proportion of the organicmaterial which is added to some soils survives in the form of a dark-colouredamorphous substance called humus. Humus is composed principally of the degradation products of the cellulose and lignin derived from plant remains. The accumulation of humus in soils is furthered by conditions unfavourable to the oxidative decomposition of organic matter. The decomposition of organic matter in bogs, ponds, swamps, and water-logged soils under conditions which are largely if not entirely anaerobic results in the production of relatively large quantities of humus, frequently of the type which is called peat. Where the organic matter supplied is distinctly acid, as under coniferous forests, or heath plants, humus usually accumulates as a definite layer at the surface. Decomposition of the organic remains under these conditions is largely effected by fungi. In prairie and steppe regions, where a grassland vegetation is predominant, humus usually accumulates in considerable quantities as a result of the decay of both underground and aerial organs of plants.

The amount of humus which accumulates in any soil depends upon the relative rates of the addition of organic residues and of its disappearance as a result of oxidation. Soils with a low humus content may result from a sparse contribution of organic matter, as in desert or semi-desert regions, or from a rapid oxidation of organic material which prevents accumulation of humus even when the supply of organic residues to the soil is large. This latter condition obtains in many tropical and sub-tropical soils. Humus is essentially colloidal in its properties and possesses an even greater imbibitional capacity than the colloidal clay particles of the soil. The plasticity and cohesiveness of the colloidal organic matter, while considerably less than that of the colloidal clay, is much greater than that of the non-colloidal fractions of the soil. Humus is rather inert chemically, its influence on the soil being largely a physical one. Its presence in soils favours a looser structure and hence better aeration, a higher water holding capacity and a reduction in cohesiveness as compared with heavy clay soils.

Since the clay fraction and the humus are the two essentially colloidal fractions of the soil, and are associated in an intimate physical relationship many soil scientists refer to these two soil fractions, considered jointly, as the colloidal complex of the soil. A large part of this colloidal complex occurs in many soils, as films enveloping the larger soil particles. The relation of these films to the maintenance of the crumb structure of the soil has already been mentioned. Many other important soil properties are either due to or greatly influenced by the colloidal complex.

SOIL WATER AND THE SOIL SOLUTION

Water is universally a component of soils, although the amount present may vary from the merest trace to a quantity sufficient to saturate the soil, i.e. completely fill all of the spaces between the soil particles. Dissolved in the soil water are varying quantities of numerous chemical compounds. These originate principally from the dissolution or chemical weathering of the rock particles, from the decomposition of organic matter, from the activities of microorganisms, and from reactions between the roots of plants and the soil constituents. It is thus more accurate to speak of the soil solution than of the soil water, although in discussions of the water relations of soils it is often a common practice to disregard the presence of solutes in the soil water. The concentration of the soil solution in any given soil varies with the proportion of water present. While in most soils the soil solution is very dilute, in saline and alkaline soils it maybe so concentrated that only a few species of plants can survive with their roots in contact with it. The principal cations found in the soil solution are Ca_{11}, Mg_{11}, K1, Na1, Al^{+1+}, and Fe^{111} (or Fe^{++}); the principal anions are HCO_3Th $P0_4$, Cl, $N0_3$, $S0_4$, and 5i03. Other solutes too numerous to mention may also occur in the soil solution but are usually present only in very low concentrations. Soil water relations receive a more detailed discussion later in this chapter.

THE SOIL ATMOSPHERE

The irregularity of the soil particles in size shape, and arrangement insures the existence of a certain amount of space between them, even in the most tightly packed soils. This is termed the pore space of a soil. The pore space of soils varies from approximately 30 per cent of the volume of the soil in sandy soils to about so per cent of the volume in clay soils, or even higher in soils which are very rich in organic matter. The pore space of any given soil depends upon the physical and chemical conditions to which the soil is subjected. Conditions favouring a crumb structure of a soil, for example, usually result in an increase in its pore space. The interstitial spaces of a soil may be occupied entirely by air, as indesiccated soils, entirely by water, as in saturated soils, or, as is most usually true, partly by water and partly by air. The relative proportions of water and air present in any given soil vary depending upon the water content of the soil.

The air present in the soil is generally referred to as the soil atmosphere.The soil air usually contains a higher concentration of carbon dioxide and a lower concentration of oxygen than the above ground atmosphere. Concentrations of carbon dioxide as high as 5 per cent have been recorded for the soil atmosphere; such values are far in excess of the average value of 0.03 per cent for the air. The accumulation of carbon dioxide in soils is due more to the metabolic activities of micro-organisms than to respiration of soil animals and the underground portions of vascular plants. Except in very dry soils the soil atmosphere is usually saturated or nearly so with water-vapour.5. Soil Organisms. The soil flora includes bacteria, fungi, and algae. The bacteria are generally the most abundant of all the living organisms present in any soil. Among them are the nitrifying, sulfofying, nitrogen fixing, ammonifying, and cellulose decomposing bacteria. The bacteria accomplishing the oxidative decomposition of cellulose and similar compounds are the most important agents in the production of humus. The numbers of bacteria present vary greatly from soil to soil, and in any one soil vary with seasonal and other fluctuations in soil conditions. Most soils contain between two million and two hundred million individual bacteria per gram of soil. The number of bacteria decreases rapidly with increasing depth, sub soils being sterile or practically so. In general an abundant representation of most species of bacteria is favoured in soils by warm temperatures (3 - 400 C.), good aeration, and a good but not superabundant water supply. A high calcium content of the soil is also favourable to the development of most species of bacteria, apparently because it favours a granular structure of the soil, thus improving aeration. Some species of bacteria, on the other hand, are anaerobic,and thrive when aeration of the soil is deficient. The denitrifying bacteria and certain of the nitrogen-fixing bacteria (Clostridiurn spp.) are examples of anaerobes.

Fungi are, in general, most abundant in soils of acid reaction. In such soils they largely replace bacteria as agents of decomposition of organic matter.The soil fauna includes protozoa, nematodes, earthworms, insects, insect larvae, and burrowing species of the higher animals. The earthworms are generally credited with having the most important effects on soil structure, at least in many soils. Their activities result principally in a general loosening of the soil, which facilitates both aeration and distribution of water. Many of the other soil animals have similar effects on the structural organization of the soil.

SOIL WATER RELATIONS UNDER FIELD CONDITIONS

In a region of moist climate, if a hole be dug or bored into the ground at almost any place where the soil is deep enough a level at which the soil is completely saturated with water will be encountered at some depth or other. Water will stand in this hole up to this level of complete saturation which is called the water table. In river valleys or in close proximity to large bodies of water the water table will usually be reached at a depth of only a few feet under the soil surface. Even in regions of arid climate there are some local situations in which a water table is present. The depth of the water table in any locality usually fluctuates, sometimes very markedly, with seasonal and periodic changes in the relative rates of rainfall, evapouration, transpiration and other factors. Relatively impermeable soil layers sometimes impede downward percolation of water sufficiently to cause the development of temporary water tables which may be far above the level of the true water table. Such temporary water tables have essentially the same effects on soil water relations as true water tables, except that their influence is often only a transient one. For many years an important role was ascribed to the capillary rise of the water from the water table into the soil above in maintaining the moisture conditions within that soil. Recent investigations have shown, however, that the importance of this source of water in soils has been overemphasized. Experiments upon the rise of water through columns of soil indicate that water seldom rises through the soil by capillarity from the water table at an appreciable rate for heights of more than a few feet. In a typical loam soil the absolute maximum capillary rise of water is about 8 feet. Such capillary rise as does occur takes place most slowly, but to the greatest height in clay soils, and most rapidly but to lesser heights in sandy soils, loam soils occupying an intermediate position.In many soils the water table lies so far below the soil surface that it has little or no effect on the soil moisture conditions in the soil layers which are penetrated by the roots of most plants. Generally speaking even in loam or clay loam soils, a capillary rise of water from a zone of complete saturation is probably ineffective in providing the roots of most species of plants with any considerable part of the water which they absorb unless the water table is within about 15 feet of the surface. Some of the more deeply rooted species may obtain some water from a water table located at depths as great as about 30 feet, but the presence of a water table at greater depths than this is an egligible factor in supplying the roots of any species of plants with water.

A capillary rise of water from a water table is usually an important source of water for plants in such locations as river bottom lands or in fields or forests in close proximity to ponds or lakes. In many and perhaps most agricultural soils in moist climate regions plants obtain some water from the water table at least during the earlier part of the growing season. However the widespread practice of drainage of agricultural regions has often resulted in such a marked lowering of the water table as to greatly reduce the possibility that capillary rise of water from below will supply crop plants with any significant part of the water which they absorb.

The other important source of soil water is that part of the precipitation usually rain, although often melting snow or ice, which percolates downward into the soil. In dry regions this is the only source of water available to plants. Even in more humid regions this is the principal or only source of available soil water in most soils during the dry season of the year. Not all of the rain which falls on the surface of a soil penetrates into it. Some maybe lost by surface run-off and a portion usually evapourates before it can sink into the soil. The proportion of the precipitation which is lost in the run-off is in general much less with open, porous soils such as those found in most forests than with those of a lower porosity. A smaller proportion of the water is lost in the run-off when it falls as a gentle rain than when it comes as sudden downpour. The proportion of the precipitation which is lost by evapouration from the surface layers of the soil depends in part on the colour, texture porosity, and other properties of the soil, and in part on the temperature and vapour pressure of the atmosphere, as well as upon the impinging solar radiation.

Let us now consider how percolating rainwater becomes distributed in as oil. We will assume that the soil under consideration is approximately air dry, that it is essentially homogeneous to a depth of several feet, and that Let us now assume that another two inches of rain falls on this same soil before any appreciable amount is lost from its upper layers by evapouration or transpiration. After a new equilibrium has been attained the upper foot of soil will not have increased in water content, but a layer of the soil approximately two feet in depth will now be moistened up to its field capacity. In other words addition of a second increment of water equal in volume to the first does not increase the water content of the already moist layer of the soil above its field capacity, but results, due to further capillary movement, in raising to its field capacity a second layer of soil, lying directly under the zone which had been moistened by the addition of the first increment of rain, and approximately equal to it in thickness. Although the exact mechanics of the distribution of water under these conditions is probably more complicated, the effect produced is as if the second increment of water simply flows through the moist blanket of soil, after which it is distributed to the dry soil layer underneath by capillarity. Successive rainfalls would thus continue to deepen the layer of soil which has been moistened up to its field capacity. If the water table is not too deep, and if sufficient rainfall penetrates the soil, not too much of which is utilized by plants growing thereon, eventually the entire soil from its surface down to the water table will be moistened up to its field capacity. A zone several feet in height just above the water table may be enriched somewhat above field capacity by upward capillary movement from the water table. If additional water is applied to the soil, after the entire soil mass down to the water table has attained its field capacity, this water percolates down through the soil under the influence of gravity and becomes apart of the ground water. Such conditions obtain in many of the soils of more humid regions at least during the wetter seasons of the year, provided that the water table is not located at too great a depth. This downward percolation of water to the water table is an important factor in influencing its depth below the soil surface, which usually fluctuates considerably from season to season.

In the preceding discussion we have assumed, in the interests of simplicity a homogeneous soil. However, most soils consist of a vertical succession of several distinct horizons, each with more or less distinctive physical and chemical properties. Even after long continued disturbance of a soil by plowing or other cultural activities at least some semblance of the original soil stratification usually persists. In the horizons below those reached by the plow the original structural organization of the soil is usually maintained practically intact. Although the individual horizons of a soil are often fairly homogeneous within themselves, each such horizon in a given soil may have a distinct field capacity of its own. The water contents of the different horizons, even after an equilibrium in the capillary distribution of water has been attained, may therefore be very different. A somewhat erratic distribution of rainfall to an underlying soil often results from cracks or channels which are opened into the soil by one agency or another. Many soils crack upon drying, sometimes to depths of several feet. Such cracks often facilitate a mass flow of rainwater to a considerable depth in the soil. Similarly in forest soils the decomposition of roots may leave channels through which water flows down into the deeper layers of the soil. The burrows of animals may also facilitate the entry of water into a soil.The presence of rock strata, or of relatively impervious layers of soil close to the surface, will also modify very markedly the simple picture which has been presented of soil water conditions.

LABORATORY MEASUREMENTS OF SOIL WATER RELATIONS

The problem of the water relations of soils has been approached both from the standpoint of laboratory and field studies. Some of the broader generalizations resulting from field studies have been just described. Of the numerous laboratory measurements of the water relations of soils which have been devised, only a few of the better known will be discussed.

Water Content. The water content of a soil is commonly expressed as the percentage of water present in terms of the dry weight of the soil. It is measured by drying a sample of the soil in an oven to constant weight at (usually) 105^{0}C. and assuming that the loss in weight represents water. There is however nothing critical about the temperature 105^{0}C. The relation between the water content as determined by this method and the temperature at which the soil is dried is a linear one over a wide range of temperatures extending on both sides of this point. This shows that there is no especial significance to be attached to the amount of water which is lost from a soil when dried at this temperature. There is also a small error in such determinations due to some decomposition of organic matter at this temperature.

Determinations of the total water content of the soil contribute very little to an understanding of the absorption of water from the soil by plants. As will be shown later the amount of water available to plants may be very different in two soils of identical water contents.

WATER RETAINING CAPACITY

One of the most familiar laboratory determinations of soil water is that of the so-called water retaining capacity (also called moisture holding capacity, water holding capacity, water capacity,etc.). This determination is usually made as follows: A shallow cylindrical pan with a perforated bottom is filled with dry soil, immersed in water to the depth of a few millimeters, and allowed to stand until it becomes saturated. The soil is then allowed to drain free of all the water which will drip out under the influence of the pull of gravity. The water content of the soil after it has been subjected to this treatment, determined by the usual method of oven-drying, is regarded as its water retaining capacity.

It has often been assumed that a determination such as that just described is an indication of the amount of water which would be held by that soil in the field in opposition to the pull of gravity but such an assumption is entirely unjustified. In the first place in making such determinations the vessel is filled with the soil in such a way that the distribution of particles in the soil mass is very different from that which obtains in the field. This error is sometimes minimized by cutting out a core of soil in the field by means of a steel cylinder and using this core for a determination of the water retaining the plant. Since no water is added to the soil the plants eventually pass into a state of permanent wilting due to the loss of water by transpiration. As soon as this occurs a sample of the soil is removed and its water content is determined by the oven-drying method.

Prior to extensive determinations of wilting percentages it was supposed that plants differed very markedly in their capacity to reduce the water content of a soil. It was assumed, for example, that species which could endure drought conditions could deplete the moisture content of a soil to a lower percentage before showing permanent wilting than could those species which were soon injured or killed when subjected to drought. Extensive investigations have shown, however, that hydrophytes, mesophytes, and xerophytes all reduce the water content of a given type of soil to about the same value before the condition of permanent wilting is induced.

The value of the wilting percentage for each species was determined by calculating the ratio of the individual determination to the mean of all determinations made with that soil. The values given in this table are the average mean ratios for a number of determinations on each species.

While the wilting percentage for a given soil shows no appreciable variation when measured by means of different plants growing in it, the value varies greatly with the type of soil. The percentage of water remaining in a soil when permanent wilting of the plants growing in it occurs ranges from approximately 5 to 10 per cent for sandy loams, from about 10 to 15 per cent for silty loams, and from about 1 to 20 per cent in clay loams. According to Caidwell (1913) and Shive and Livingston (1914) a greater percentage of water is left in the soil at permanent wilting if the plants are transpiring rapidly than if they are transpiring slowly. Veihmeyer and Hendrickson however, record that the wilting percentage of a given soil as determined with sunflower plants shows very little difference over a wide range of climatic conditions. In general this quantity seems to be controlled almost entirely by soil conditions and is only slightly influenced by the species of plant used, or by the climatic conditions to which that plant is exposed. The significance of the wilting percentage lies in the fact that it is essentially a measure of that fraction of the soil water which is unavailable forplants. However, it should not be assumed that absolutely all movement of water into the roots has ceased when the soil water content has been reduced to its wilting percentage. Wilted plants continue to reduce the water content of the soil but at such a slow rate that restoration of turgidity is impossible since the rate of transpiration even from a wilted plant exceeds the rate of absorption of water from a soil which is at its wilting percentage or lower. Actually, plants in the field often exhibit permanent wilting at soil water contents somewhat in excess of the wilting percentage as determined by the usual technique. In laboratory determinations of the wilting percentage the plants are ordinarily rooted in relatively small containers in which the soil mass becomes effectively penetrated by roots resulting in a fairly uniform reduction in the water content of the soil. In the field, however, such conditions do not often prevail. The soil is usually less effectively penetrated by the root systems of the plant growing therein, and some portions of the soil may be less completely depleted of water than others before permanent wilting ensues. Although the soil in the immediate vicinity of the absorbing zone of each root is at the wilting percentage more remote portions of the soil may still be at the field capacity. Hence the average water content of such a soil at permanent wilting may be considerably in excess of the wilting percentage as determined by the usual laboratory procedure.

THE DIFFUSION PRESSURE DEFICIT OF THE WATER IN SOILS

In preceding discussions of the water relations of plant cells and tissues it has been shown that the most significant unit for the expression of the dynamics of water relations of plant cells is the quantity which has been termed in this book the diffusion pressure deficit. Since all problems of the absorption of water by plants involve a consideration of the relation of the water in the soil and the water in the plant, it is desirable that this concept of the diffusion pressure deficit of water be extended to the soil water if the absorption of water is to be interpreted in terms of dynamic units.

One of the first successful attempts to measure the retentive capacity of the soil for water was made by Shull. In an ingenious approach to this problem air-dry seeds of the cocklebur (Xanthium pennsylvanicurn) were used as the instrument for measuring the diffusion pressure deficit ofthe soil. Seeds of this species were first placed in salt solutions of various osmotic pressures and the percentage of water which had been imbibed when an equilibrium was attained in each solution was first determined. Once the percentages of water held by these seeds when in equilibrium with solutions of different osmotic pressures had been determined, the diffusion pressure deficit of any medium in which they might be immersed could beme asured. For example, if some of these seeds were immersed in a medium of unknown osmotic pressure (diffusion pressure deficit) until a dynamic equilibrium was established, after which it was determined that their moisture content was 32.8 per cent, a diffusion pressure deficit of 26.6 atmos. was indicated for the medium. This principle was applied to the measurement of the diffusion pressure deficit of soils at different water contents. A sample of soil at a known water content was shaken with some ofthe seeds until an equilibrium was attained, after which the water content ofthe seeds was determined.

With an increase in the soil water content its diffusion pressure deficit decreases. In the range of low water contents a small increase in water content corresponds to a very large decrease in the diffusion pressure deficit ofthe soil. As the water content of the soil is increased further the diffusion pressure deficit continues to decrease but at a progressively slower rate. Diffusion pressure deficit values for soil water may be calculated from the results of a number of other types of determinations. Among these are vapour pressure measurements determinations of the freezing point depression of the soil, and measurements of the amount of water retained by a soil in equilibrium with tensions of a known magnitude. Each of these methods is suitable only for a certain range of soil water contents. The first can be used satisfactorily only with relatively dry soils, the last with relatively wet soils, and the second only over an intermediate range of soil water contents. From data obtained by one or more of these methods it is possible to construct for any soil a water content-diffusion pressure deficit curve. Most ofthe important soil water relation scan be interpreted in terms of such curves. Such a curve is characterized by an abrupt rise in diffusion pressure deficit in the region of low soil water contents.

The position of the wilting percentage is just to the right of the sharp inflection which occurs in the curve. The diffusion pressure deficit of the soil water at the wilting percentage lies within a range of 412 atmos. Experimentally determined values for this quantity vary slightly depending uponthe method used for their measurement. The position of the wilting percentage makes understandable the fact that all species of plants seem to be able to reduce the soil water content of any given soil to approximately the same percentage before they pass into a state of permanent wilting. As shown by this curve reduction of the water content of the soil by only a very small percentage below that of the wilting percentage results in a very rapid rise in the diffusion pressure deficit of the soil. Plants can only obtain water when the diffusion pressure deficit of the soil water is less than that of the peripheral cells in the absorbing zone of the root. Except in halophytes the diffusion pressure deficit of root cells probably seldom exceeds 25 atmos. and is usually considerably less. The diffusion pressure deficit of the water in a soil will attain an equilibrium with a plant in which a diffusion pressure deficit of 25 atmos. has developed in the root cells at only a slightly lower soil water content than with a plant in which the diffusion pressure deficit of the root cells is only 5 atmos. Hence wilting percentages as determined with different species of plants are all approximately the same since the actual differences in wilting percentages usually don't exceed the experimental errors inherent in the method of determining this quantity.

To the right of the point corresponding to the wilting percentage the slope of the curve is very gradual. The field capacity corresponds to a diffusion pressure deficit of about one atmosphere, indicating that plants have very little difficulty in absorbing water at soil moisture contents corresponding to this value the preceding discussion has been based on the assumption that the soluble salt concentration of the soil is not sufficient to have an appreciable effect on the diffusion pressure deficit of the soil water. In any soil heavily charged with soluble salts, as for example alkali soils, the diffusion pressure deficitof the water may range up to 100 atmos. or even higher, due to the osmotic effect of the dissolved salts. The curve expressing the soil water content-diffusion pressure deficit relationship in such a soil would not approach very closely to the zero axis, even in soils which are nearly or entirely saturated but would level off at a value corresponding to the osmotic pressure of the soil solution.